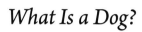

What Is a Dog?

What Is a Dog?

Raymond Coppinger and
Lorna Coppinger

Foreword by Alan M. Beck

The University of Chicago Press

Chicago and London

Raymond Coppinger is professor emeritus of biology at Hampshire College. **Lorna Coppinger** is a biologist and science writer. Their books together include *Dogs: A New Understanding of Canine Origin, Behavior, and Evolution*, also published by the University of Chicago Press.

The University of Chicago Press, Chicago 60637
The University of Chicago Press, Ltd., London
© 2016 by The University of Chicago
Foreword © 2016 by Alan M. Beck
All rights reserved. Published 2016.
Printed in the United States of America

25 24 23 22 21 20 19 18 17 16   1 2 3 4 5

ISBN-13: 978-0-226-12794-1 (cloth)
ISBN-13: 978-0-226-35900-7 (e-book)
DOI: 10.7208/chicago/9780226359007.001.0001

Library of Congress Cataloging-in-Publication Data
Coppinger, Raymond, author.
    What is a dog? / Raymond Coppinger and Lorna Coppinger ; foreword by Alan M. Beck.
        pages ; cm
Includes bibliographical references and index.
    ISBN 978-0-226-12794-1 (cloth : alk. paper) — ISBN 978-0-226-35900-7 (e-book) 1. Dogs. 2. Dogs—Behavior. 3. Human-animal relationships.
I. Coppinger, Lorna, author. II. Beck, Alan M., writer of foreword. III. Title.
    SF422.5.C66 2016
    636.7—dc23

                                                              2015031906

♾ This paper meets the requirements of ANSI/NISO Z39.48–1992 (Permanence of Paper).

We dedicate this book to
Peter Neville and all our friends at COAPE
and
to Hampshire College and all our friends there

CONTENTS

FOREWORD

The legend for the Kato, an indigenous Californian group of Native Americans, is that when the Creator set out to construct the world, he took along a dog (Leach 1961)—a wonderful metaphor in which the dog precedes not only people but every other thing as well. While for many that is not exactly correct, the dog, indeed all canids, enjoy a special role in most human cultures. A major topic addressed in *What Is a Dog?* is that the role is by no means the same through time and, especially, place.

I joined the community of dog-watching scientists, like almost all the others, from other interests. I completed my master's studies on the effects of wildfire on prairie plants and animals (rodents and deer) assuming I would move up the food chain and study wolves. My new major professor at the Johns Hopkins School of Public Health, Edwin Gould, wrote me a handwritten letter noting that while wolves are interesting, because of public health implications, studies of urban dogs are less expensive to conduct and much more fundable; in addition, there were many wolf studies, but, at that time, there were no urban dog studies, so I studied the dogs of Baltimore, Maryland, and pretended they were wolves. Over the years, I learned to appreciate that even a relatively small and isolated dog population as found on the streets of Baltimore was quite varied. There were dogs that were always under direct supervision via leash or voice commands, pets that were roaming for various periods of the day or night, and a small population that were always free to roam, either abandoned to the streets or escaped, and even a smaller group that were born on the streets, though they had a short life or were taken off the streets by animal control or by people to become house pets. They were remarkably diverse in size, shape, color patterns, and behavior. Those variations allowed me to use photo capture to identify individuals as concisely as if they were purposely marked, as done with other animal field studies. With visually marked individuals, I could study behaviors and home range and estimate population abundance without ever capturing or recapturing any animal. Because dogs are so accepting of humans as part of their environment I could watch them relatively closely without ever changing their behavior, a luxury not possible with most wild animal studies.

The study (Beck [1973] 2002), because of society's fascination with all dogs, attracted wide interest, and I learned that the easiest way to become a world's authority is to be the only one doing what you do. Early in my new dog career, I was invited to give a talk at the University of Massachusetts, where I was invited to stay over with Ray and Lorna Coppinger. The talk was in the evening so Ray suggested we use the day to go dogsledding. This was a totally new experience for me. He set me up with a four-dog sled, and I followed him with his sled with its full pack. Ray is an excellent teacher so I welcomed his training me in dogsledding, but he forgot tell me that sleds have a brake. All went well for some time, as I followed, but when Ray had to stop, I kept on going — perhaps Massachusetts's first rear-end dogsled pileup. The dogs were fine and we can only wish we knew what an idiot they thought I was.

I have shared just a little of the variety of experiences associated with our relationship with dogs — if a single small Baltimore dog population could be so varied, how can we begin to appreciate the world's dog population and, indeed, the world's canid population? *What Is a Dog?* addresses the much greater world of evolution, domestication, and global view of the canids and their role with people worldwide. As the title declares, it is not even clear just what a dog is. It is an animal that entered the world at different times and probably from different routes and now lives in many ecological niches, both human developed and wild and natural.

This amazing book is about all dogs, and of all the canids around the world — for example, wolves, coyotes, jackals, and wild dogs — 95 percent are what we call dogs. There are reasons to consider all of them as one species, and they may all have had some role in developing the animal we now call dog. The ubiquitous and popular dog of developed countries exists in large part because of the economic growth and stability of those countries and not because of some sophisticated biological or evolutionary process. In the developing world, dogs are loved and are supported by a large social commitment and great economic infrastructure of breeders, commercial products, and specialized health care. In most of the world, however, dogs are viewed as somewhat useful at best but more often as pests in conflict with human needs and health. McKnight (1961) noted that "the dog is a domestic animal and in an exodomestic situation is at best pitiful and at worst destructive and dangerous." *What Is a Dog?* explores and analyzes that view for a better and more complete understanding of the roles of dogs in different places around the world.

In this book, the Coppingers look at the world's dog population not by species or even by geographical categories but by local social roles: restricted dogs (fully dependent and fully restricted); family dogs (fully dependent and semirestricted); neighborhood dogs (semidependent and semi- or unrestricted); and feral dogs (independent and unrestricted). In this way, the book tackles the fascinating question: Are the village dogs around the world essentially different than the many breeds that populate our homes and dog shows in the developed world? The reason the planet can support the huge dog population, both pet and pest, is because human beings provided all dogs with a place and sometimes food and protection, indeed a niche found around world — a niche in which dogs have lived for thousands of years. But without people, the village dog and pampered pet alike would soon disappear — dogs would become extinct.

What Is a Dog? explores this view by an assessment of the food energies necessary for the individual for acquiring food, scavenging, reproduction, protection from dangers, especially raising young, and even repair costs after injury from competition with others of the different dog populations. This insightful analysis explains how the different canids evolved and how the newest canid, the domestic dog, can survive in their new niche.

The Coppingers remind us that "the overwhelming majority of people on earth do not think of dogs as companions, to be owned and paid for," but at the same time, much of the world enjoys dogs as companions, indeed, as members of the family to be protected and loved. This understanding is best achieved by knowing just what a dog is.

Alan M. Beck

References

Beck, A. M. (1973) 2002. *The Ecology of Stray Dogs: A Study of Free-Ranging Urban Animals.* West Lafayette, IN: Purdue University Press.

Leach, M. 1961. *God Had a Dog.* New Brunswick, NJ: Rutgers University Press.

McKnight, T. 1961. "A Survey of Feral Livestock in California." *Yearbook of the Association of Pacific Coast Geographers* 23:28–42.

PREFACE

In the fall of 1970, Hampshire College — a brand new four-year liberal arts college — opened its doors. Its quest was for a new approach to teaching undergraduates, with classes being closely combined with original research and field-work. We were hired partly because we were active with working dogs. The reasoning was that many of the students were interested in dogs, thus dogs would be the "bait" with which to attract students to science. Ray and his students were racing sled dogs and had them "wired up" to find out how they could be the fastest animal in the world for marathon distances. Lorna was writing *The World of Sled Dogs*.

By 1977 the sled dog book was published and our dog-sledding days were numbered. The climate was changing, and snow was becoming hard to find around Amherst, Massachusetts. However, in the mid-1960s we had begun to read reports of a little wolf-like animal appearing in the forests. Since we had been out there running dogs, we also studied what then was known as coydogs, a supposed cross between domestic dogs and western coyotes.

What was it? What a great question it was for developing young science students. Here was a big mammal and the big scientists were discussing what it might be.

The debate about what species it is has been going on ever since. Helenette and Walter Silver of the New Hampshire Fish and Game Department first thought it might be a cross between a coyote and a dog. Then in 1969 Barbara Lawrence and Bill Bossert at Harvard measured the skulls and compared them to wolf and coyote skulls at the Harvard Museum of Comparative Zoology. The skulls were neither like wolves nor quite like coyotes but were closer to coyotes than wolves. Therefore, they were officially classified as coyotes. Those who were not happy with this just called them "the New England Canid."

Our students began to ask the question, "How did they know the skulls in the museum were wolves or coyotes?" Well, because, whoever wrote the tag for the skull wrote they were a wolf or a coyote.

Not a great answer. A Hampshire student, Michael Sands, published a paper in the *Journal of Mammalogy* in 1976 in which he compared the histology and behavior of the so-called coydogs' sweat glands with those of wolves in Alaska and coyotes in the Southwest and he concluded the animal was a wolf. That wasn't a happy finding because the wolf was on the endangered species list, which meant the management of such an animal would fall under federal law and the state wildlife specialist wouldn't be able to manage them.

Even now, with DNA research supplanting the measurement of skulls and the identification of blood type, scientists still cannot say definitively what this animal is. It's been labeled a coydog, a coywolf, an Algonquin wolf, and an eastern coyote — but perhaps it is its own animal. What a wonderful predicament for our students to study: an animal the experts don't know what to call even after spending nearly half a century trying to figure it out. The "answer" was not important for our class; our engagement with hypotheses and measurements were. Our students were out there doing their thesis work and publishing papers.

While we were racing dogs, another phenomenon took place. In 1973, the urban ecologist Alan Beck published *The Ecology of Stray Dogs*. The book represented an abundance of firsts, including being the first to look closely at street dogs as valid research subjects. Up until then, they were known as mongrels or mutts—ubiquitous strays with no planned breeding behind them. Beck's study of these free-ranging dogs used the methodologies of ecologists. He used the same techniques that were being used by field biologists who studied wolves or coyotes and he asked the same kinds of questions—how many animals were there and were these animals earning a living on the streets of Baltimore?

This wasn't exactly what we had been working on, but it was about dogs and we were interested. We invited Alan to Amherst where he was to give a joint talk to the Five College Community. Hampshire hosted students and faculty from Smith, Mount Holyoke, and Amherst Colleges and the University of Massachusetts.

Alan was given an exciting dogsled ride through cold and snow-packed Wendell State Forest as a reward.

In late 1977, the Winthrop Rockefeller Foundation (thanks to William Dietel of the Rockefeller Brothers Fund) and Hampshire College (thanks to its then-president Charles Longsworth) awarded us a challenge grant to investigate the potential of Old World livestock-guarding dogs as livestock protectors in the United States. Laws in the United States were changing about shooting, trapping, and poisoning wildlife blamed for livestock depredation. If the guarding dogs had been successful for over two thousand years in the Old World, couldn't they be successful here? Our project was active from 1977 until the mid-1990s, as we explored the sheep/goat/cattle pastures of Europe and the Middle East. We found "rare" sheepdogs and ran field studies to understand how they worked. We learned amazing things about dog behavior.

As our explorations expanded, we were astonished to see that many of the dogs that were part of the sheep/shepherd-dog combination in many countries looked very much like Alan's stray dogs of Baltimore. People thought of the dogs as a breed—their own special breed. Then in South Africa, we were introduced to Edith and Johan Gallant. They had just finished writing a book on the African dog, which they called affectionately Africanis. The frequent claim is that the dog of the Zulus and other native peoples of Africa is their own special breed of African dog.

The Gallants taught us about the specialness of these archetypical dogs. All of a sudden these ubiquitous dogs captured our attention, and we were off on our third big dog adventure.

For starters, Italo Costa at the Canadian Broadcasting Corporation's *Nature of Things* series asked if we would write a script for an hour-long documentary about the world's working dogs. Of course!— and do it up right and write something spectacular. The script was a dog person's dream but surely nothing anybody could afford to do. But we were wrong, and Ray and Italo ended up traveling—filming Maasai cattle herds, monkey hunts in Pemba, sheep dogs in Italy and Saskatchewan, and Border collie trials in Scotland.

The TV show *Nova* followed and asked for a similar program. The producers were restricted in budget, and we ended up filming village dogs in the Tijuana (in Baja California) dump. The Tijuana dump was a beautiful place—almost psychedelic. Colored plastic blowing in the wind—thousands of seagulls—and *hundreds* of dogs. We were fascinated. A few years later we shifted to the Mexico City dump and started some serious studies. We found a wonderful population of dump workers who wanted to help us.

One time, during a lecture in New York State, Sue Sternberg (founder of Rondout Valley Animals for Adoption) asked if we could teach a course in the dump. Arie Kopelman of Puppyworks, the dog training and behaviorist company, organized it, and the registrations poured in—from people who wanted to go to a huge dump and look at dogs. And then Jane Brackman (canine science writer) provided some support money for Alessia Ortolani (our former student who had worked with us on the dogs of the mountain villages of Venezuela) to do a study on the village dogs of Ethiopia. We were especially interested in the relationship of village dogs to the endangered Ethiopian wolves. While there, Alessia found a behavioral difference between the dogs in Christian and Muslim cities and towns. She published her work in the scientific journals and went on to work with our Italian colleague Luigi Boitani, who with his students and collaborators has done some of the best studies of the behavioral ecology of village and free-ranging dogs.

And so our studies took us to the dumps: collections of vast amounts of some of the waste generated by humans every day. In all our study areas, we found hundreds and hundreds of dogs that are regarded as mongrels, strays, and the cast-offs of the domestication process. For us, behavioral ecologists, the free-ranging dogs that forage, reproduce, and live without human guidance, turn out to be the most thought-provoking dogs we have studied so far.

PART I

About Dogs

1

What Is a Dog?

It's a *really* good question: what is a dog? This question echoes the one asked by our paleontology professor Albert Wood at Amherst College many years ago. His question was "What, if anything, is a rabbit?" Was a rabbit a rodent, descended from the ancestors of species like squirrels and rats? Perhaps it was a hyrax, a little shrew-like animal thought to be related to the elephants, or perhaps the rabbit descended from the ancient marsupials such as ancestors of kangaroos. Somebody thought it was closely related to the primates, which is still a reasonably good hypothesis. The discussion, which started long before Professor Wood's 1957 paper, continues yet. Still nobody quite knows what a rabbit is. All the while, we students knew exactly what a rabbit was when we saw one.

Years later, with two colleagues at Hampshire College, we emulated Professor Wood's classy title and wrote a book chapter called "What, If Anything, Is a Wolf?" The answer to this question is important for wildlife biologists and lawmakers who need concrete evidence on which to base decisions about whether an animal is an endangered species — for example, is it a gray wolf, a red wolf, a coyote, or perhaps a hybrid between them — or maybe even a dog? What is a dog?

Currently, there is a suggestion, call it a movement, that the gray wolf (*Canis lupus*) be removed from the endangered species list because, as the argument goes, its numbers are high and it occupies all its former range. Well, what was its former range? Is the former range measured from the last glacial period, or when the Pilgrims reached America, or from the beginning of the twentieth century when the American government decided to eliminate all wolves from the lower forty-eight states? The sticky question is that government trappers also killed off red wolves and Mexican wolves — and are those different species or even a different subspecies from gray wolves? People argue in court over such designations.

Did the gray wolves ever raise pups in New England and, if so, are they repopulating this former range now? Some argue they never were in New England. How do you know that? Or maybe wolves live in New England now but we have been calling them coyotes by mistake. Some scientists think the New England canid is a wolf — not the gray wolf but a different species of wolf called the Algonquin wolf. At present, the Algonquin wolf is not on the endangered species list because nobody knew they existed as a separate species.

Still other people continue to think the big coyote-like animal living in New England is a cross between a coyote and a dog — a coydog, but others think it is a coywolf. Well, if they are coywolves, as hybrids they are not protected under the Endangered Species Act. Thus our question is still debated among scientists: "What, if anything, is a wolf?" It is important to know, and we still don't.

We grew up with all these different species and thought we knew what they were. Our moms read to us about the big bad wolf in "The Three Little Pigs," and the youngsters watched Wile E. Coyote cartoons on Saturday morning TV. And then there was *Lassie* and *Rin Tin Tin*.

Dogs are ubiquitous in most of the world. It is hard to imagine anyone on earth not having been exposed to a dog at one time or another. We all know a dog when we see one.

Within the dog family (the Canidae)—coyotes, wolves of all different species and subspecies, jackals of all different species and subspecies, and dingoes, as well as all the species of foxes, dholes, and bush dogs—the dog differs from all the rest of them in many ways, as you will read in this book. Within the genus *Canis*, wolves, coyotes, jackals, and dingoes are generalized predators. We call them the wild types, to distinguish them from domestic dogs, *Canis familiaris*, which we call simply "the dog." The wild types look and behave like predators. Scientists study the phylogeny—that is, the evolutionary history—of this closely related genus; they want to know which are more closely related to which, and which evolved first, and when did they evolve, and where and how did the divergence (speciation) between them take place.

With the domestic dog, those questions of when, where, and how they evolved are almost like an obsession for many of us. The dog is different, special, a beautiful animal, and thus we want to know more about the process that produced such an unusual species. How did such an animal evolve so quickly?

Many think *C. familiaris* is really another form (or subspecies) of the wolf. And maybe it is. But of which wolf species are we thinking? Many who tackle the question are comparing the dog with the large canid creature along the Last Glacial Maximum (about 25,000 years ago) when they look for doglike deviations in the gray wolf skulls. Professor Wood, however, used to waggle his finger at us and say, "If you want to know anything about the evolution of a species don't study their skulls."

When some of us say "wolf," the mental image forms along the lines of "I know a wolf when I see one." And it is big, not a little jackal-sized wolf.

We are in a time when "everybody" claims the dog is a subspecies of the gray wolf. For those of us who are scientists, that is an open question and not an answer. "Method of thinking" is another way of saying that logic, statistics, and science have rules. When one takes a statistics course, one learns the rules of what statistics can do and what it cannot do. A statistician comparing measurements between animals that are labeled "dog" and "wolf" might say the probability that dogs descended from the wolf is $p < .001$. That means the chance of the relationship being due to chance alone is less than one in a thousand. And most of us would then say that is a fairly high chance that dogs descended from the wolf.

Similarly, the logician might say that:

all wolves have forty-two teeth of a specific arrangement;
that specific number and arrangement are also found
    in the dog; and,
therefore, dogs and wolves are the same species.

It isn't as silly as it sounds because that is exactly the way the anatomist or systematist thinks. That was the kind of argument that our clever Professor Wood applied to try to find the origins of the rabbit. He counted teeth and looked at the shape of the teeth, and yes, paleontologists looked at the angulation and attachment of the jaw and other organs in the body to get clues as to the origin of the rabbit. Linnaeus counted teeth, which is one reason why he put dogs, wolves, coyotes, jackals, and dingoes in the same genus, *Canis*.

For both the statistician and the logician, the conclusions aren't necessarily right or wrong. But in each case the method has its limits in what it can tell you.

Science is just the same. Scientists test hypotheses. Dogs and wolves are related, is the hypothesis. Is it true that only wolves and dogs have forty-two teeth? No: jackals and coyotes and dingoes also have forty-two teeth. Then look for other differences between them. They are all, for example, interfertile—maybe they are all the same species. So why do the taxonomists assign them into different species?

The naming of wolves, dogs, coyotes, jackals, and dingoes was all just an unfortunate consequence of history. In 1758, Swedish biologist Carl Linnaeus named the dog *Canis familiaris*. That was a hundred years before Darwin published a theory of evolution. Almost everyone in the early nineteenth century, including scientists, believed that God had created the different species just as they are. Coyotes, dingoes, and the whole lot of wolves and jackals were given their binomial name (i.e., genus and species names) by missionaries, explorers, artists, and other travelers—not knowing that in 1858 the definition of species was going to change.

After Darwin, the definition of species was very simple for evolutionary biologists: a species is a sexually isolated population of animals (or plants). Because wolves, coyotes, jackals, dingoes, and dogs are not only interfertile but regularly interbreed and hybridize, by definition they are all the same species. That should be the end of the story.

But of course, it is not. When did dogs diverge from wolves, if they did? Recent scientific papers have concluded that the divergence occurred 130,000 years ago or 30,000 years ago or 15,000 years ago. The correct answer is none of the above. Biologists in the nineteenth century classified every geographic variation as a different species (including people). It was okay to enslave some peoples because they were a different species. In fact, the dog has not diverged from the wolves, coyotes, or jackals or their relatives, according to modern evolutionary theory.

In addition, a statistician might ask the question a different way: how long ago did wolves, dogs, coyotes, and jackals share a common ancestor? Joseph T. Chang, a statistician at Yale University, worked out that solution for people. He argued that everybody has two parents and four grandparents and eight great-grandparents (fig. 1). By the time you get to ten generations you have a thousand ancestors, and at

How Many Great Great Great Grandparents Does One Have?

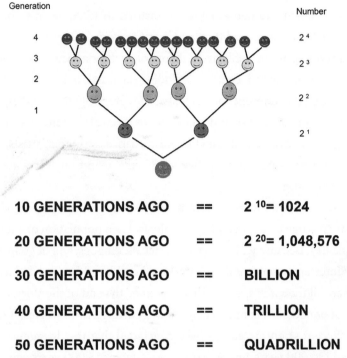

10 GENERATIONS AGO	==	2^{10}= 1024
20 GENERATIONS AGO	==	2^{20}= 1,048,576
30 GENERATIONS AGO	==	BILLION
40 GENERATIONS AGO	==	TRILLION
50 GENERATIONS AGO	==	QUADRILLION

Figure 1. Sexually dimorphic species all have two parents and four grandparents and so on back for generations. A dog, in a hundred years or fifty generations, could have had a quadrillion great-great-great-grandparents. This seems impossible since there have never been a quadrillion dogs. The only way to explain that fact is that all dogs had to share great-great-grand-grandparents with all other dogs and, probably, wolves, coyotes, dingoes, and jackals.

twenty you have a million, and at thirty, a billion, and at forty, a trillion, and at fifty, a quadrillion.

Well, there never has been a quadrillion people — so there were not enough people to make you, or alternatively, we all had to share ancestors. When was the time when everybody on the planet had the same ancestors? Fifty-four generations ago, everybody on the whole planet had the same grandparents.

That might be a thousand years ago. That might mean that with a twenty-year generation time, everybody in South Africa and everybody in Japan (or name your favorite spot) could trace their ancestry to William the Conqueror. If Marco Polo had left a surviving lineage (not everybody does or did), then everybody on the planet today could claim him as a direct ancestor.

Suppose we asked the politically incorrect question: Did white people evolve from black people or was it the other way around, and did they both evolve from Asian types? The argument would begin to sound absurd, as well it should.

If we asked the same question for the members of the genus *Canis* isn't it equally absurd? If fifty-four generations ago is the magic number for people, why isn't it the same for dogs? The fifty-four generations ago for the *Canis* (races) is 200 years. Two hundred years ago all jackals, wolves, dingoes, coyotes, and dogs shared the same ancestors.

Do we believe it? It is the not the job of a scientist to believe something but rather to test the hypothesis with the scientific method. Chang applied a brilliant statistical technique for figuring out the time to the most recent common ancestor. Does the method tell you when the various races of humans appeared on the earth? No. Is when they appeared the important question? No. Does the method tell when the first dog appeared on earth? No. Does it tell you how the first dog evolved? No. Is that an important question in understanding dogs? No.

Maybe the statisticians are wrong. Have they overlooked something? Is Chang's reasoning faulty? Chang's conclusion is such a spectacular way to look at the evolution of a species. It is fascinating. Our hypothesis starts, however, with the assumption that dogs, coyotes, and the rest are the same species. According to the original definition, that is true. And yes, by definition they are totally interfertile. (So for this chapter they are all the same species — maybe we will change our minds later.)

Historically, scientists classified dogs, wolves, jackals, and all into different species because of dozens of reasons — such as they live on

Figure 2. A wolf in northern Quebec and a dog on the East African island of Pemba (*top left* and *top right*, respectively). The skull at *bottom left* is an adult wolf; the one at *bottom right* looks like a dog but is a yearling wolf. (Skulls from the Museum of Comparative Zoology, Harvard University. Skull photos by Abby Drake.)

different continents or some are bigger or smaller or some have proportionately longer muzzles than others. Wolves have longer snouts than dogs—unless you are comparing wolves to borzois. Figure 2 (*bottom*) shows two canid skulls side by side. The one on the left is a wolf. The one on the right is proportionately shorter and wider, and those are traits that archeologists look for when they are looking for the first dog. But the skull on the right is not a dog. It is a juvenile wolf. You can tell that because the sutures in the juvenile wolf's palate are not all closed, which means the animal was still growing and changing.

You have to compare wolves to the average dog. What, then, is an average dog? And so on and so on. Speciation studies often measure and compare skulls. Yet our professor advised us to stay away from skulls when trying to delineate a species. Single skulls found here and there don't tell what the total variation of the population is. They grow at different rates according to the environment, showing seasonal or annual differences, as do the rings of a tree. Jackals and coyotes have very similar skulls but live continents apart. They occupy similar niches on those continents, and their skull morphology has converged on the best shape for doing the job.

Scientists need to create a better hypothesis. That is what scientists do. They test and retest hypotheses. If the data support a hypothesis, great—test it another way and see if you can falsify the hypothesis. For it to be science, it must be repeatable. It is mandatory to try to repeat it. If the data do not support the hypothesis, then change the hypothesis and test again.

Philosophically, scientists can never say they have found the truth. Some of us in the dog world are irritated by statements that scientists have found the truth about the origin of dogs—as if a single origin was identifiable. Look! We found the first dogs, the oldest dogs, the Adam-and-Eve dogs. That is truly silly. The search must be for a population of dogs. At that point you have to ask the question, What is a dog, anyway? What are we searching for?

Supposedly, we all know what a dog is. Someone holds up a puppy and says, "What is that?" and you answer, "That's a dog."

Could you ever be fooled? Maybe. Maybe it is only part dog. Years ago we ran a team of sled dogs, and often when buying a dog we were informed by the seller that this dog is one-eighth wolf. We heard it so many times that we filed the statement under "myth," or "factoid." We became suspicious that a purported wolf ancestor is an advertising gimmick. We were expected to think that if the dog has a little wolf in it, it would be a better sled dog. Knowing what we know now, we should have said, "If it has a recent hybrid wolf ancestry, then we don't want it!" Why would anyone want any physical or behavioral characteristics of wolves on a sled-dog team? It would be trouble all the way. The sled dog—not the wolf—is the animal selected to pull sleds far and fast. Biologist Erik Zimen once put a "sled dog" team of hand-raised wolves together to pull him on a sled. He reports a host of hilarious happenings in his book *The Wolf, a Species in Danger*.

But there seems to be an appeal for a dog to be part wolf. One hears or reads about the wolf-like breeds. Statements abound that German shepherd dogs look more like wolves than other breeds or that Siberian huskies are an ancient breed, implying that both German shepherds and huskies are more closely related to wolves than are other dogs. Or perhaps they are descended directly from wolves as a breed.

Those are interesting but false observations because we know just when the German shepherd (at one time called the Alsatian wolf dog in Great Britain) became German shepherds and what breeds of dogs Max von Stephanitz crossed together, in the late 1890s, creating the fine sheepdog called the German shepherd. Imagine creating a wolf dog to herd sheep.

It is also known that Siberian huskies became Siberian huskies as a registered breed in the 1930s at Chinook Kennels in New Hampshire.

Similarly, the Anatolian shepherds were registered with the American Kennel Club (AKC) in 1996, using some of the dogs we had collected in rural Turkey as part of our livestock-guarding dog project a few years earlier. These are not "ancient breeds."

Most purebred dogs like the German shepherds, the Anatolian shepherds, the Siberian huskies, the Border collies, or the golden retrievers can be traced back to one or two or a handful of founding stock—all within the past 150 years. Those of us who are interested in the genetics of breeds know that if we take a sample of genes from purebred dogs, those genes couldn't possibly represent the diversity shown in the population of dogs from which the sample was chosen a hundred years ago. The Siberians and the shepherds and the golden retrievers were created by crossbreeding with other breeds. The makeup of those breeds has been carefully recorded. Which breeds make up our modern breeds is not a mystery.

And how successful they have been! In the United States now there are as many golden retrievers as there are wolves in the whole world.

Because they looked so uniquely doglike and different from every other species, the dog was named *Canis familiaris* by Carl Linnaeus in 1758. In his pioneering binomial system of classifying all the plants and animals, the dog was designated as its own species, unique in the animal/plant world. In the 1980s, systematists changed the name of the dog to *Canis lupus familiaris*, claiming thereby that the dog is not only a descendant of the wolf but really is a wolf. Oh, what a terrible mistake they made.

Even if all dogs were descendants of "the" wolf, that doesn't make it a wolf or even a subspecies of wolf. Some say the dog can mate with the wolf and produce viable puppies, and that is true. They also can and do mate with coyotes, Ethiopian wolves, three different species of jackals, and dingoes and produce viable puppies with all those species. What makes the gray wolf so special?

Geneticists claim that dogs and wolves have similar gene arrangements and that the two are more like each other than they are to other members of the genus. Then they claim that red wolves are a cross between coyotes and gray wolves, that the New England coyote is really a wolf hybrid, and that the coyote has dog genes as do Ethiopian wolves and so on. It doesn't seem, however, that geneticists can produce repeatable results.

Whatever the dog is descended from, some of us think and are prepared to argue, with data, that the dog has crossed the boundary between it and its ancestor and that it is its own species within the taxonomic family called Canidae and the genus *Canis*. In this, we side with

Linnaeus and against the modern systematists. (See: we have already changed our minds.)

The dog is a different size and a different shape and behaves very differently from any of its supposed ancestors. Figure 2 (*top*) shows the relative size and shape of a gray wolf and a dog. Many people look at the wolf and think—wow, it is beautifully adapted to being a wolf: it hunts big animals like elk and moose.

Then they will look at the picture of the dog in figure 2 and say, "Well, it isn't a real dog." (For fun, one could search through pictures of all the dogs in the world and see which picture fit the image of "the real dog" the best.) The consensus is likely something along the lines of: "The pictured dog is some kind of stray dog or a mongrel. It is a mutt, or a cur, a kind of pavement special, a pariah, or maybe even a feral dog."

If you obtained that dog from an animal shelter you first might name it Spot, and then try to determine what breeds contribute to its looks. Send its DNA off to a lab, and they will send a report of which breeds they detected. Perhaps Spot has some beagle genes, or Jack Russell terrier? What were the parents?

The picture of our Spot (fig. 2, *top*) was taken on the island of Pemba, off the coast of East Africa in the Indian Ocean. No official kennel club breeds of dogs existed on Pemba when we took this photo. Very possibly, since the beginning of dogs there have been no purebred dogs on Pemba. Since evidence of kennel club activity is nonexistent there, Spot cannot be a crossbred mongrel.

It is a religious taboo to keep pet dogs on Pemba and thus Spot can't be a stray dog—the product of irresponsible ownership.

We will make the argument that Spot and the millions and millions of free-ranging dogs all over the world are the real dogs. We should be studying them because these "village dogs" or "street dogs" like Spot contain the essential essence of dog. They are the pervasive dogs of the world. When we refer to dogs in this book, we mean village dogs, unless otherwise noted.

Aristotle defined a species as an essence. There is an essence, an emanation of its spirit, a revelation of its core being, to any species, be they wolves, pigeons, or dogs. Biologists and philosophers have discussed the fine points of what it means to have the essence of a species. When Saint Thomas Aquinas talked about that essence, did he mean the same thing that Darwin or Aristotle did? Probably. Whatever they understood, their mental process was no doubt akin to what takes place when many of us identify great art: we all know what an essence is when we see one.

Pigeons have an essence. In a way, pigeons, with their pigeon essence, are very much like dogs. Most pigeons, but not all, are unrestrained street or park pigeons that are people oriented but still essentially a "wild" animal. Some places in the world are loaded with pigeons and in other places none exist. They find food around the city or village and then fly off and raise their squabs. Some cities have enormous populations. In Istanbul, Milan, and in Trafalgar Square or in Central Park, the pigeons look essentially alike. Are those pigeons, sitting on Admiral Nelson's head, strays, mongrels, street pigeons? Are they feral pigeons? Are they perhaps pariah pigeons? Are they domestic pigeons? Are they domesticated pigeons gone wild (fig. 3, *top*)? These are all the same questions people ask about dogs.

The number of pigeon breeds (813) exceeds that of dogs by almost two to one. Some of them are working pigeons, carrying messages. Famous war pigeons, with names, were heroes and received medals for their bravery. A pigeon by the name of Cher Ami saved Galloping Charlie Whittlesey and the lost battalion in World War I. Some of these pigeons have been stuffed and displayed in national museums.

Sporting breeds of pigeons are bred for pigeon racing. Dozens of pigeon clubs organize championships for local, state, and national races. At a pigeon hall of fame, you can see pictures of winning pigeons. You can buy books on how to raise and train pigeons.

A few of the exotic breeds have bizarre looks and behaviors: flat-faced breeds with feathers growing on their toes or growing in the wrong direction; big pigeons and pygmy pigeons. At breed shows

Figure 3. *Top*, Stray pigeons or maybe feral pigeons or maybe pet pigeons in just about any park. *Bottom*, Dogs around the world also tend to be uniform in size and shape like the pigeons. Trained hunting dogs in East Africa, pictured here, look like any other village dog.

and at arena performances, rollers and tumblers—different types of performing pigeons—compete for looks and behavior. People spend months training and grooming and preparing their pigeons for the show ring. Commercial pigeon foods, purchased by the bag, are guaranteed to give your favorite show pigeon a glossy coat. Some people keep pigeons as pets. People build pigeon lofts that are fancy kennels for pigeons where they can be sexually isolated and purposefully bred.

Fanciers write about purebred pigeons, racing pigeons, and working pigeons in picture and pigeon health magazines. Leaflets and fliers announce shows and races and other pigeon events. Agencies handle lost pigeons.

It seems that some people eat pigeons (squab are a delicacy), and other people have them as pets and couldn't imagine eating one.

And certainly there are people like Ray's mother who, in an upscale town in Florida, fed the wild, feral, stray, mongrel, street pigeons. She had a pigeon feeder on her lawn and she would put food in it every day around 2 P.M. Fifty pigeons would show up at 1:45 and sit on the neighbor's roof, turning it white while waiting for mom to show up with their food. She was always excited because they would be in a frenzy all around her before she even put the commercially bought pigeon food into the feeder. She referred to them as her pets. We also know of an old lady in Istanbul who sold pigeon food in the square outside the Blue Mosque so the locals and tourists could feed the pigeons and take pictures of little kids with pigeons sitting on their heads.

That all sounds a lot like the dog world. In fact, one could substitute the word "dog" for the word "pigeon" above and it would read okay, except that dogs have fewer breeds (451, some say). So why don't people call the thousands of pigeons in all the parks around the world strays, mongrels, street pigeons, feral pigeons? Why don't the humane societies have sterilization programs to reduce the populations of pigeons?

Aren't the vast majority of pigeons in the world—like the vast majority of dogs—beyond the reproductive control of humans? Aren't 99 percent of the world's pigeons responsible for finding their own food and raising their own squabs? We should call most of the pigeons in the world village pigeons and city pigeons or street pigeons.

Why don't we ask, "When did man first domesticate the pigeon?" Why don't we see statements such as, "The pigeon was the first domesticated bird"? Why don't we say the pigeon is the most variable bird in the world? Have people domesticated the pigeon, and for what purpose? When did pictures of pigeons first appear on temple walls and pottery? Maybe a better question would be, why would you care when, where, or how people first domesticated pigeons? In fact it is fairly obvious that humans didn't domesticate pigeons. It looks like pigeons domesticated themselves when the new niche called "the city" evolved. So why do we think humans domesticated chickens, cats, or dogs? Is it because they are a different shape, color, and behavior than the wild ones? Is it because they live around people?

There is no question that dogs are a domestic animal, just as pigeons are a domestic animal. "Domestic" simply refers to animals that live around or with or beside people or simply in the presence of people: the domestic dog, the domestic rat, the domestic pigeon, and the domestic chicken. Also, there is the ubiquitous house sparrow (*Passer domesticus*), which lives not only in cities and villages but also on farms. The house mouse, cockroach, and bedbug are all more common around humans than anywhere else. All in some sense are referred to as domestic — the domestic cockroach!

No one would think that humans intentionally domesticated the rat, cockroach, house mouse, or house sparrow. It is hard to imagine that people domesticated the pigeon. So why do we believe that people domesticated the dog?

Domestic does not mean domesticated. Domesticated has a different meaning. Domesticated means someone or some group of people intentionally and selectively bred a wild species to be domestic (related to the home environment). Someone captures wild pigeons, sexually isolates them by keeping them in cages, and breeds them for exotic behaviors such as rolling and tumbling in flight.

So it is with dogs. Dogs running around the world's villages vary from black to white. Someone captures a couple of white dogs from a village population and locks them in a kennel, thus sexually isolating them from the larger population. These breeders let their white dogs breed only with other white dogs. At this point they become a breed with a breed standard — they must be white. Breeds are the results of breeders and breed clubs controlling the dogs' reproduction. Dogs (or pigeons) don't exist as breeds in the wild. In the wild, regional variations of a domestic animal are known as landraces. A landrace is very much the same as a subspecies — a geographically based population within a species that contains a nonrandom distribution of alleles (forms of a gene).

That white breed of dogs is a domesticated dog because someone intentionally isolated them and bred only white dogs to white dogs. It does not mean that when you go into the village and see a white dog, that it is a stray dog escaped from sexual isolation. It doesn't mean that any dog that doesn't belong to a sexually isolated breed is a mongrel. It also doesn't mean that all the nonwhite dogs are mongrels. And especially it doesn't mean that the original dog was white.

A person or group picks white as a recognized breed characteristic. Another group finds beagle-colored dogs on the upper Nile River, an indigenous dog that the locals use for hunting. According to breed-book legend, Swiss hounds (which became the backbone of many of the European hunting breeds) were brought from the Nile River to Europe by the ancient Phoenicians and developed as a breed — domesticated in Switzerland. If true, that does not mean that if I go to the Nile River and see a dog of the same color that this was the ancestor of our modern hunting dogs, nor does it mean that hunting dogs evolved on the Nile River, or that the original dogs were domesticated to be hunting partners of people.

Spot (fig. 2, *top*), is typical of most of the dogs seen around the world. Spot weighs a tad less than thirty pounds. The wolf in the picture spends its summers on the tundra and its winters in the forest of northern Quebec. This one weighs about seventy-five pounds or maybe a little more, after we fed it a salami sandwich.

Spot and this wolf differ in almost every characteristic, the most obvious being size, color, and ear carriage. But the biggest difference is their behavior. They have different breeding cycles and they differ in their parental behaviors. Certainly the dog in every respect differs more from the wolf than the wolf differs from the coyote, jackals, and other wild types.

At some time in the past, a wolf-like population of animals changed from the wild shape to the domestic shape. Changed, to an evolutionary biologist, means a change in the gene frequency of a population of animals. It does not mean that a pet wolf changed into a dog. In order for any evolutionary process to occur, there has to be an evolving population.

This book is about all the dogs in the world. It is important to understand that many people who read a dog book often think of "dogs" as kennel club creations. It is the purebred dog that is man's best friend, not the street dog. This book is not about purebreds or about "man's best friend" (a distressingly overused and possibly totally untrue phrase). Man's best friends are storybook dogs. Man's best friends live ubiquitously in the United States, Europe, and other developed countries and, in these countries, are by and large household pets. Man's best friends only live in areas where people have easy access to vaccines against rabies and distemper. They are the results of certain levels of commercial appeal involving pet stores, breeders, dog food companies, veterinary medicine, magazines, and books.

This isn't to deny that a large part of the world's dogs are pets. We need to add our professional opinion, however, and to display a little squeamishness toward the fancy hobby pet group about the few hundred Western breeds that set the standard for what dogs are supposed to be.

Could it be that breeds represented as working, or hunting, or pet groups don't represent real dogs? Could it be that the so-called stray dogs, street dogs, neighborhood dogs, village dogs, and even feral dogs of the world are the real, naturally evolved, self-selected dogs?

We will argue that those street dogs are not mongrels or strays. We will argue that they are the real dogs, the ancestral type of our modern breeds. They are unique and beautifully designed by evolution.

Whether they are on the streets or in people's yards or houses or in the dump at the edge of town or if they only come out at night, they are part of a continuous worldwide and ancient population of dogs. They are much more ancient than any "ancient" breed.

For the moment, we are going to declare that Carl Linnaeus got it right. The dog is its own, individual species. A lot of us think it is a beautiful species. And not because it looks or acts like a wolf but just the opposite: it doesn't look or act like a wolf—it looks and acts like a dog.

We are going to test "What, if anything, is a dog?" in a different way in this book. Instead of using the methodologies of paleontologists and geneticists, we thought we would try the methodologies of behavioral ecologists.

Behavioral ecologists are scientists who study populations of animals. A number of our colleagues study the reintroduced wolf population at Yellowstone National Park. As a reintroduction program, it was intended to establish a population that would grow in that environment. How do the wolves in Yellowstone earn a living? What do they eat, and how much do they eat? How many puppies do they have each year, and what is the survival rate of those puppies? How long do they live?

In part, what is interesting here is that ecologists define species in a different way than do evolutionary biologists, paleontologists, geneticists, or anthropologists. The ecologist defines a species by its niche. "One species, one niche" is their credo. We are going to put our behavioral ecologist hats on and ask ecological questions about dogs. This is a different approach to dog studies than most people are familiar with, but we think it will answer the question: What is a dog? We think our story will demonstrate that dogs are animals just like any other species and have to obey the rules of behavioral ecology. Just because they are easily adoptable and because they can be bred and trained to perform certain tasks doesn't mean they are fundamentally different as a species.

That is our hypothesis, as we investigate what, if anything, is a dog.

2

The World Is Full of Village Dogs

The number of dogs in the world is phenomenal. Of the seven (or so) species in the genus *Canis*, domestic dogs are 95 percent of them. But these dogs are not necessarily house pets. The vast majority of dogs in the world run their own feeding and reproductive lives.

What *is* the number of dogs in the world? Instead of saying almost a billion, let us say it *is* a billion. That is a better number because it easier to divide by than "almost." These are large populations, and as populations dispersed all over the world, they are very hard to count.

Also, populations of animals can change dramatically from minute to minute. If today there were a billion adult dogs, half a billion of them would be females. If dogs were like wolves (which they aren't) and all the females had five pups on April 15, then on April 16 there would be 2.5 billion day-old puppies. Add the billion adults, and the total is 3.5 billion dogs. That is a staggering number.

The first rule in counting any animal population is to

determine whether the number given includes all age groups. Surely someone is going to criticize these figures and say that a billion dogs is too high and that the number is probably half that. But if half of the 500 million are females (250 million females), giving birth to five puppies per year amounts to well over a billion puppies.

When the experts count they arrive at an estimate. When the estimate is given, a margin of error is included—a billion dogs, say, with a factor of two. In other words, would you believe the population could be as high as 2 billion (2×1 billion) or as low as half a billion ($\frac{1}{2} \times 1$ billion)? For some estimates, a factor of ten would be good enough. For dogs, for example, if you count puppies you could be off by a factor of 3.5 ($1 \times \frac{1}{2} \times 5 = 2.5 + 1 = 3.5$).

The other problem, of course, is who is included in the count. When the U.S. Census Bureau counts humans, the count is inclusive of all age groups from newborns to adults. The newborn puppies are hard to count, however, and the question is whether they count in any meaningful way. Certainly, if it was said that there were 3 billion dogs including newborn puppies, the expert would be skeptical because that doesn't exactly represent a stable population. In other words, if all puppies survived, then our dog population would be growing at over 2 billion dogs a year, and it wouldn't be very many years before the entire planet would be ten feet deep in dogs. For the population to stay stable, most of the puppies have to die.

Sometimes when the experts are counting a species, they report the number of animals per square kilometer. They take a finite area and count really carefully. Then they estimate the total number of kilometers in the dog world and multiply to come up with the big number. If we wanted to know how many dogs were in Sri Lanka, we would sample certain areas and then simply multiply that number by the total area of Sri Lanka. Then you might assume that the density of dogs there was the same in other countries such as South Africa. However, the density of dogs varies enormously not only from place to place but from time to time. But still these figures are useful. They reveal a hypothetical population at a specific point, which is useful for management purposes. If we were going vaccinate all the dogs in Zimbabwe, we would need to know how many injections to take with us.

So how do you count dogs in the whole world? Do you organize hundreds of people in many places around the world to go out at the same time on the same day to count dogs, then collect all the figures and add them up? Believe it or not, such a system is used in several countries for counting birds. Here in the United States, the National Audubon Society sponsors the Christmas Bird Count. Volunteer ornithologists and bird-watchers go out on a designated day in a designated area and count how many birds of each species fly by and at what time of day. Then the interested people collect the numbers from the field-workers and put them all together for an annual report. Why do they do it at Christmas time? Because in winter, the trees have no leaves and thus birds are easier to see and because fall migration is over and so there's no need to worry about nestlings.

How accurate are their final numbers? The final number is, obviously, just a guess; call it an educated guess. It turns out that the final number is not the important number. The important finding is how this year's number compares with last year's and with the numbers in the years before that. While there's no illusion of accuracy, the numbers do provide a good idea of whether the populations of different species are rising or falling. During the fifty years we have lived on our property, for example, the towhee population has dropped every year. It might be time to put them on the endangered species list and find out what's going on. Similarly with yearly estimates for any endangered species, it's possible to keep tabs on population swings. If you are counting over a large area you can look for geographic changes.

At our house we have a battery of bird feeders, each one designed to attract a different species of bird. For example, we have a glass milk bottle hung horizontally, and only the chickadees can get through the mouth of the bottle to the sunflower seeds within. At intervals, we have to refill the bottles; the rate at which we have to refill reflects the number of birds feeding from those bottles. We started that system years ago for an interesting reason. We wanted to try to guess how many chickadees were coming to our feeder — maybe fifteen or twenty? One morning we decided to capture the birds with traditional nylon mist nets and band them so we could tell them apart. At 250 birds we were late for school and had to drop the nets. In the weeks that followed, we could compare the number of banded birds to the number of unbanded birds at the feeder and get an estimate of the total population. Doing a rough calculation, if half the birds coming to the feeder were banded, then we had a total of 430 birds. This is called the mark-recapture technique. But in our case, the number of banded birds was much less than half. Instead, we had close to a thousand chickadees per day. It is easier to count seeds. In a normal year we replace an estimated 4,000 sunflower seeds a day. If the number of seeds gone in a day changes, then we assume the population has changed.

Alan Beck at Purdue University might have been the first to use the mark-recapture technique on dogs using a camera. For his PhD thesis, published in 1973 as *Ecology of Stray Dogs: A Study of Free-Ranging Urban Animals*, he took pictures of dogs on a street corner or a dump at a certain time. Then at a later time he took another picture at the same place and looked at the proportion of new dogs to previous dogs in the second picture and then the third and so on. Then one just has to do the math. Using this mark-recapture technique, Alan estimated the street dog population in Baltimore at 50,000 dogs in 1970. It would be interesting if Alan would go back now for more pictures. Wonderful cameras nowadays will automatically take pictures of all passersby, including dogs. Video cameras can collect similar data.

Methods for counting populations such as dogs are labor intensive, time consuming, and expensive (fig. 4, *bottom*). They are dependent on dogs' activity levels, reproductive activity, and dozens of other variables. We were always suspicious, during many of our dog studies, that there was an active nighttime population of dogs that we could never get a handle on. In Italy, biologist Luigi Boitani once mounted lights around a dump, and sometime in the middle of the night he would turn the lights on and count the wolves, dogs, and other critters that were feeding during those late hours.

One question about the accuracy of the estimates is, Why do you want to know the number? Most of the time, the actual number of dogs is often just a curiosity: "Gee whiz, there are a billion dogs in the world!" A scientist might be curious or interested in devising a method to solve a new problem in measurement. Statisticians have come to fame for devising clever techniques such as the mark-recapture method. One of our favorites, of course, is the time-honored count-their-legs-and-divide-by-four routine.

If you are looking at human health issues on which dogs can have an impact, such as the spread of rabies, you would be interested in the number of dogs vaccinated compared to the total number of dogs in a region where rabies is present. Or you might need to know how much vaccine to take to an area where rabies is prevalent. You might be interested in the dog-bite problem and want to know how many bites per month per dog.

Knowing the number of dogs in a population isn't as important as knowing what the age structure of the population is. How many new dogs would you have to vaccinate every year, assuming that once a dog is vaccinated it will not need another shot in its short life span?

We did such a study on a working dog population of livestock-guarding dogs (fig. 4, *bottom*). What we wanted to know was how many puppies would have be to raised each year in order to keep the population stable at a thousand working dogs? What we found out was that most livestock-guarding dogs are not really effective until they are about two years old. That is true of many working dogs. Our sled dogs weren't matured enough to be top dogs until at least two years old, and wolves aren't proficient hunters until they are at least two.

Here is the problem. For us, the job was to keep a thousand dogs between the ages of two and seven working. Starting with a thousand dogs, we found that, on unfenced ranches, 50 percent would be dead

Dog Restraining Practices around the World

% Unrestrained	Location	Source
20-36%	Ecuador	WHO (1988)
33%	USA	Beck (1975)
68-72%	Italy	Boitani & Racana (1984)
72%	Sri Lanka	WHO (1988)
78%	Zambia	De Balough et al. (1993)
86%	Tunisia	WHO (1988)
96%	Sri Lanka	Matter et al. (2000)
99%	Zimbabwe	Butler & Bingham (2000)

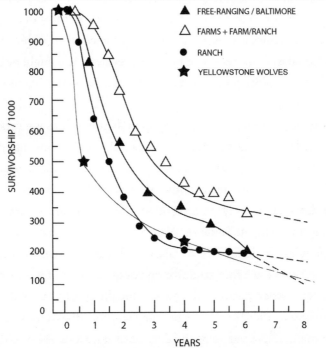

Figure 4. *Top*, Dog-restraining practices around the world from the WHO.
Bottom, Age structure of three different populations of dogs and Yellowstone wolves.
(Adapted from Alan M. Beck, *The Ecology of Stray Dogs* [1973], Jay Lorenz et al.,
"Causes and Economic Effects of Mortality in Livestock Guarding Dogs" [1986],
and personal communication from Douglas Smith.)

by eighteen months of age or before they were two. On farms, 50 percent were dead by thirty-three months. Thus every eighteen months or thirty-three months, there would have to be 500 replacement dogs. That means that on ranches one needs to have just under 400 new pups every year. On farms, the number is somewhere between 250 and 300 per year just to keep the total population at a thousand. But that isn't what we wanted. We wanted a thousand dogs between the ages of two and seven years, thus the number of replacement pups needed to be a good bit more (fig. 4, *bottom*).

If we were working with wolves and had about the same expectations we'd have to ask the next question: Is the carrying capacity of the niche a thousand adults (two to seven years old) with their pups? That isn't a good question. A better one would be: What is the number of pounds of wolves that the niche can support? In other words, how many pounds of wolves between two and seven years old are needed to support how many pounds of a reproductively stable population? The total number of pounds of wolves is determined by the size of the niche.

If we were veterinarians and we found that the life expectancy of dogs in the Mexico City dump was similar to that of the ranch dog, the replacement rate would be 500 new dogs per 1,000 dogs every year and a half. That isn't quite right because the dump dogs have a shorter life span, to perhaps their fifth birthdays, in contrast to the livestock dogs, which are effective until their eighth birthdays (seven-plus years old). Thus, to maintain a population of 700 adult dump dogs, the replacement rate of dogs reaching their first birthday has got to be 140. That is the population that needs to be vaccinated. The other 400 pups won't make it to the vaccination date.

Age structure of dog populations varies over the surface of the earth. The age structure of the dogs in the Mexico City dump is distributed differently than in most wild populations of wild types. In the dump, almost half the dogs were suckling puppies, and the other almost half were adults. Juveniles (the age class between suckling pups and reproductive adults) were just a tiny percentage of the total. In Ethiopia, Alessia Ortolani found an age structure that showed similar results, with juveniles being less than 10 percent of the total population. Comparable findings in central Italy by Luigi Boitani and colleagues led them to think that, in spite of the puppies born, little successful reproduction was taking place. If you were running a vaccination program, maybe all you needed to vaccinate (and mark) was the adult population because the number of dogs surviving their first year was very low.

In the 1990s we were working on a book in which we estimated the number of dogs in the world. The hypothesis was that the greatest density of dogs was at the equator and the least dense place was the North Pole. Pick a country halfway between those two places for which there is an estimate of its dog population (Italy), multiply that estimate by the total number of countries in the world (200), and the total comes to 400 million dogs.

That calculation seemed reasonable at the time, but being careful scientists, we checked it with a different method. We counted dogs in east African villages and South American villages and came up with the figure of seven adult dogs per one hundred people. At the time, the world's human population had just reached 6 billion people. If you did the math, it would come to 425 million dogs. We were happy with having arrived at essentially the same answer in two different ways and published the figure.

Consider a margin of error with a factor of two. How did we feel at that time about a population being 800 million? If puppies were included, that figure was within bounds but sounded to us at the time a little high. How did we feel about the dog population being half that, or 200 million? That seemed too low. If, as in the earlier example, all the females had five pups, then the population would have tripled on April 16. Or perhaps a disease had reduced the population of dogs in Italy prior to the count, which would have lowered the worldwide estimate. Just thinking of Italy as an average country might raise a few eyebrows. None of us who love Italy ever thinks of Italy as an average country!

Almost twenty years later, we were writing a paper with colleagues and again needed to know how many dogs were in the world. Andrew Rowan of the Humane Society of the United States shared his extensive data with us. He had estimated rather carefully the total number of dogs in quite a few countries and cities around the world. He also had collected sample numbers from other researchers on specific areas.

Still there were many areas where there were no figures. So we estimated those places by figuring the number of dogs per hundred people in a neighboring country or region and extrapolating to the unknown population. If you knew how many dogs per one hundred people there were in Ethiopia (which we kind of did), then one could just find out the number of people in neighboring Somalia. Were we concerned that, Somalia and Ethiopia being ecologically very different from one another, the estimates might not be accurate? Of course we were.

Yes, these methods are crude. When we filled in all the blanks, we came up just short of a billion dogs and rounded up. Bringing the factor of two into play again would result in a total of 2 billion dogs, or one dog for every 3.5 people in the world. That must be too high. How about our old number—about 500 million, or seven dogs per hundred people? That seems more likely, but in other places the estimates are so much higher. In the townships in South Africa, for instance, we counted ten dogs per hundred. That is 40 percent more than the estimate of seven dogs per hundred people. The United States has one dog for every five people, which if typical would mean 1.4 billion dogs in the world. But the United States, like Italy, could never be considered an average place. So much for estimating by extrapolating—there is no average place.

We think the world dog population is a billion dogs, and we are sticking with that estimate. Go ahead—prove us wrong. (But that's a scary figure—fourteen dogs per hundred people? That *must* be too high? And again, what about puppies?)

When dealing with dog populations, one faces another sticky problem. The number that really matters is: How many dogs do we have reproductive control over?

If we were estimating the population of rats, regardless of the reason, the assumption is that they are a population of animals that forages and reproduces on their own, outside human control.

Why don't we think about dogs that way?

Aren't dogs just like any other animal? We have seen pet goldfish that someone has bred to have exotic characteristics. They have long tails and fancy flowing fins. They live in a bowl, sexually isolated from the world's wild goldfish. So now we are going out into the world to look for populations of goldfish that live natural lives. Have they escaped from irresponsible people like us? Are they mongrel goldfish because they don't have billowy tails? Are they stray goldfish? It might be possible! Could you think of the alligators living in city sewers after people flushed them down toilets as products of irresponsible ownership? Probably not a true story.

The problem with these free-breeding populations of dogs, cats, rats, or what have you, running around leading natural lives, is they are often pests. With many pest species, we try to control their populations by limiting their food source, interfering with their reproduction, or simply killing them. Whether blowflies, mosquitoes, cockroaches, rats, or something else, the way to control their population numbers is to clean up the mess that they are feeding on and put out the proverbial rat poison.

A billion dogs is an enormous population, and it appears that many or most of these dogs have control over their own feeding resources and their own reproduction. They are what the United Nations World Health Organization calls unrestrained dogs. In places like Sri Lanka, 96 percent of the dogs are unrestrained, and the population was estimated at 3,700 dogs per square kilometer (fig. 4, *top*).

Again, estimating the total numbers is difficult. What is most significant is that, among the billion dogs, all but a few are not reproductively controlled by humans.

Of the 70 million dogs in America, how many are spayed or neutered? Certainly sterilized dogs are under our reproductive control. But the report is that half of the litters born in the United States are unplanned. Realize that means human control over dog reproduction is far from 100 percent effective; the dogs are doing what comes naturally.

If half the litters born elsewhere in the whole world were unplanned, then there would be something like a quarter of a billion unplanned litters a year. But most countries in the world have less reproductive control over dogs than we do. In the United States, Canada, western Europe, Japan, Australia, and perhaps a few other small but atypical places, attitudes are different about controlling dog reproduction. These countries are unusual in the sense that they contain a high percentage of people who think humans are supposed to control dog reproduction. In other words, it is a very small portion of the world's population of people that thinks reproductive control over dogs is an option.

On the matter of being concerned about controlling dog populations, realize the difference between people and governments. Most governments, including our own here in the United States, see dogs as pests and problems. A talk about dogs with selectmen in our town is greeted with dismay. Some claim that half the issues that come before their weekly council meetings are dog problems. Every week, reports are made of barking dogs, nuisance dogs, dog bites, and dogs knocking over trash barrels, chasing bicycles, nipping joggers, and on and on.

In much of the world, dogs are also a health problem. Rabies is particularly serious. Seventy thousand people every year die of that awful disease. In the South African province of KwaZulu-Natal, 46,000 people per year require treatment for rabies because they were bitten by dogs. Think of the cost. Many countries need international assistance in dealing with dog problems. The United Nations has a dog agency that just follows such issues.

When governments try cheap and Draconian solutions to dog "overpopulation," people often protest. When we ask our Muslim friends, who don't really like dogs and don't keep pets, why they don't cull their burgeoning population of street dogs, or even sick and diseased dogs, they answer, "We can't. The dogs belong to God." Belonging to God or not, when an epidemic of a horrific disease breaks out, armies are called in to shoot all the dogs.

Kathryn Lord compared the reproductive activities of the species in the genus *Canis*. She and her colleagues, reviewing the literature, estimated that 17–24 percent of the dogs in the world were sexually restrained. Those figures seem conservative, and it seems to us that, of the billion dogs in the world, humans have reproductive control over only 15 percent of them. That means that 850 million dogs—domestic dogs—are well placed to run their own reproductive lives.

One other remarkable fact about that huge number of dogs: if you count all the coyotes, all the different species of wolves, all the different species of jackals, and all the dingoes, they amount to about 5 percent of the dog population. If you think the dog descended from the northern gray wolf (*Canis lupus lupus*), then the comparison in numbers gets to the gee-whiz level. The current estimate is 400,000 gray wolves left in the world. That means the wolves are 0.04 percent of the dogs—less than four hundredths of 1 percent. Dogs are classified (technically) as wolves (wrongly, we think), *Canis lupus familiaris*, which would mean that the subspecies, dog, is 99.96 percent of the species wolf (*Canis lupus*). Say it another way: dogs outnumber wolves by 2,500 to one.

That makes naming the dog *Canis lupus familiaris* look a little silly. But if it is true that dogs descended from gray wolves, then they are evolutionarily very, very successful.

3

Why Do Village Dogs All Look Alike?

Of the billion dogs in the world, three-quarters of them look as much alike as do the individuals of any other species.

A few years ago we asked a Navajo shepherd what a Navajo sheepdog looked like. He said, "A Navajo sheepdog is not too big and not too small." To us the Navajo sheepdogs were identical in size and shape and color variations with the sheepdogs of Sonora and the village dogs in the mountains of Venezuela or the ones we worked with in eastern and South Africa or saw in India and China.

That is true of the majority of dogs in the world — they are not too big and they are not too small. One of the most fascinating details about that 85 percent of the dogs in the world that control their own reproductive life is: they all look alike.

The similarity between the pigeon world and dog world continues. Pigeons, in some sense, all look alike. The pigeons in the Mexico City dump fly and look just like the pigeons in Trafalgar Square, like the pigeons in Istanbul, like the pigeons in Central Park, like the pigeons in Milan. Wherever you go, the pigeons in the park look like the pigeons in every other park.

No two pigeons are the same, of course. No two pigeons are exactly the same color or size or shape. At the same time, they all look pigeon-like. They have an essence that evokes pigeons. "I know one when I see one."

It is true for every species. The chickadees at our feeders all look very much alike, and it takes practice to see the little differences that distinguish them. They all can get into the bottle feeder as far as we know. The same is true with our blue jays and squirrels. The squirrels are intriguing because around here you sometimes see a black one or a brown one, but it still looks like a squirrel. At one time we lived on a small island of nesting seagulls. After a few years, we could distinguish the boys from the girls because of subtle differences in their head shapes.

All wolves look alike. But the wolves also show small variations of a neutral monotone. It could be that wolves vary more in coat color than squirrels do. Thus, a pack of gray wolves (*Canis lupus*) might have mostly gray animals except for an anomalous white one or black one. In some regions, the wolves are mostly white. Right now there seems to be an increase in frequency of black wolves in Yellowstone National Park. Color variations appear in certain subspecies of the gray wolf: the red wolf (*Canis lupus rufus*), the Arctic wolf (*Canis lupus arctos*), and the Mexican wolf (*Canis lupus baileyi*). Yet even with a color difference, they still look like wolves. (Taxonomists are confused about which scientific name to give to some of these wolves, but they recognize the essence of wolf in all the variations.)

In any given area, the wolves tend to be the same size. From the far north to east of the Mediterranean, they will grade in size from larger to smaller. A biologist would say this gradation is a cline, that the species follows Bergmann's rule: it grades from a large animal in the north to a smaller size on the equator. Ecologist Val Geist once pointed out that the cline isn't always perfect. Wild sheep also exhibit these clines. For example, the bighorn sheep (140–300 pounds, 3–3.5 feet) of the

northern Canadian Rocky Mountains grade down to the smaller mou-
flon of the Iranian desert (90–120 pounds, up to 2–4 feet at shoulder
height) with smooth coats. What's noteworthy is that all those different
"species" of sheep along the cline are interfertile from north to south—
including domestic sheep. As with the dogs, the spot on the cline from
which the domestic sheep evolved is difficult to pinpoint. The big gray
wolves in the north are interfertile with all the other members of the
genus, all the way to little jackals in equatorial Africa. The genus *Canis*
appears to us to be a single species cline.

Free-ranging street and village dogs, also, tend to be bigger regionally in the north and up into mountains, and smaller in equatorial regions. In Greenland, on Baffin Island, and over in the Hudson Bay area,
the village dogs we have observed can weigh as much as sixty pounds,
whereas equatorial dogs are basenji-like and weigh less than twenty-
five pounds. With increasing latitude and altitude, dogs tend toward
being rough coated.

So, if the village dogs range from twenty to sixty pounds and from
smooth coats to rough, how could we say they all look alike? It is a
good question. For us, the population density of dogs weighs heavily
on our thinking. The farther you get from the equator, or the higher
in the mountains, the fewer the street or village dogs. In the warm climates, the density can be substantial. When we want to study village
dogs, our preference is to go south (toward the equator) rather than
north. Those regional warm-weather dogs, all about the thirty-pound,
lion-colored variety, are usually prevalent. This strongly indicates that
the overall size and color of the local dog is an adaptation to the local
geography, the climate, and the prey base—in other words, the niche
in which they make their living.

Every once in a while we will see a report that scientists have discovered a new species of mammal. That means they have discovered a new
shape in a population of animals that are sexually isolated from all other
species (well, maybe sexually isolated, but not always). They name it
with a Latin binomial indicating the genus and species. It might not be

a bad system if the biologists stuck to the rules. Many people contend that dogs and wolves are the same species, that is, the dog (*Canis lupus familiaris*) is a subspecies of the wolf (*Canis lupus lupus*). The classification of any species should be mostly about biology/evolution but it can also be about beliefs, culture, politics, and numerous other factors. When a wild canid was first discovered in New Hampshire in 1944, after a lot of talk and measurements it was classified as a coyote (*Canis latrans*). The animals were bigger than the well-known western coyote (also of course *Canis latrans*). Barbara Lawrence and William Bossert at Harvard measured skulls of wolves, coyotes, and the New England canid and concluded that the New England animal was, although not exactly the same as coyotes, closer physically to them than to wolves. Well, those of us who had studied with Professor Wood did not believe skull measurements to be accurate indicators of species, and our research, done later with our student, Michael Sands, revealed histologically (shape) that the sweat glands in their feet resembled those of gray wolves in Alaska and not those of western coyotes, although few people seem ever to have read that paper, published in the *Journal of Mammalogy* in 1976.

It was always suspicious that had the New England canid been classified as a gray wolf it would have fallen under the Endangered Species Act. That would have led to any number of management restrictions about how fish and game scientists in New England states could manage this population. We now have lots of these wild canids in this region. The discussion is heating up as the U.S. Fish and Wildlife Service decides whether it is time to take the gray wolf off the Endangered Species list because its population is increasing. If the eastern coyote is really a gray wolf, then it is not rare and not endangered. Again it looks like whatever species it is, the discussion is more political than scientific.

Interested people have debated the ancestry of the dog since the late 1800s. Not only have wolves, jackals, or dingoes been suggested as the ancestor of dogs, but several people argued that dogs were the result of hybridization between wolves and jackals. There were astonishing theories about big dogs (breeds) evolving from the Chinese wolves. The Nobel Prize–winner Konrad Lorenz at one point suggested that some "breeds" of dogs descended from wolves and others from golden jackals. When we met him in 1978, he started the conversation by saying, "Everything I have written about dogs is wrong—but it was better that I discovered it rather than someone else."

In 1982, the systematists reclassified the dog (*Canis familiaris*) as a subspecies of the gray wolf (*Canis lupus*). Its scientific name became *Canis lupus familiaris*. The claim was that various wolf skulls had shorter muzzles and looked as if they could be dogs. One author found several skulls in a museum in Belgium that were classified as wolves, but she and her colleagues wrote instead that the skulls must be the first dogs from about 30,000 years ago. She didn't have the same paleontology professor we did, and her skull measuring results evoked criticisms.

More recently, the claim is that genetic evidence supports the hypothesis that dogs were descended from the wolf and only the wolf. But with every new paper on the subject, conclusions vary about when, where, and how the transformation from wild type to domestic dog occurred.

Our argument is a little different. We are studying all those dogs that look alike and are outside human reproductive control. There never was a "first dog." If a gradual transition occurred from the wild types to the domestic dog, then when and where do you draw the line between them? The Adam and Eve story of a caveman taking a wolf pup from the den and raising it to be "domesticated" might be symbolic of how domestication came about, but it is not very useful for an evolutionary biologist or a population geneticist. What happens in evolution is a progression of a new shape, in this case a dog shape, as the animal slowly adapts to a new niche. The evolved dog shape and size is unique. Whenever or wherever the beginning was, a population of thirty-pound animals evolved within a new niche in the same way that wolves evolved from their antecedents in their niche to be wolves. This evolving new animal developed the size, shape, and colors that suited its survival in the new environment.

However that ancestor/descendant discussion plays out, the dogs on a Korean city street look like the street dogs in India, like the village dogs in Africa, like the dogs in the Mexico City dump. They all show the essence of dog, which differs remarkably from any species of wolves, jackals, coyotes, or dingoes. Aristotle and Linnaeus had it right—they are a different species. They are shapes and behaviors that are different one from the other and adapted to their own niche. Dogs, as we noted earlier, are more different from the rest of the genus than the wild types are from each other.

The 85 percent of the dogs called village dogs or street dogs around the world that are outside of human reproductive control are as much alike as the pigeons around the world are alike. Alessia Ortolani, in an extensive study of the village dogs in the Bale Mountains in Ethiopian villages, sends the same report as our Navajo friend: "You never find a big one and you never find a little one." And in the pictures she sent us, the dogs look just like the dogs we were studying in the Mexico City dump, in KwaZulu-Natal, in a Chinese dog market, in Vietnam, and in the Navajo Nation (plates 1 and 2).

Here is the question: Do the village dogs around the world have a different essence than all those breeds of dogs at the AKC? When watching the dogs in the Mexico City dump, a number of the students would say, "These dogs are different from real dogs—these are mongrels." In other words, the implication is that the kennel club breeds are the ancestors of the village dogs.

People seem to believe that if a dog doesn't look like one of the kennel club recognized breeds then it must be a hybrid or mongrel. People think if you let all the pure breeds go and they interbreed for a few generations, the resulting population of dogs would look like the Mexico City dump dogs (plate 3). However, this cannot be true. For instance, the dogs on Pemba all look the same, and they have never been anywhere near any breed of dog.

Breeds of dogs are a result of what Darwin labeled artificial selection, which he contrasted with natural selection. He was impressed with, even in awe of, people who created breeds of pigeons or dogs or the myriad of farmyard species. Darwinian artificial selection happens when humans select for a particular trait by sexually isolating a male and a female from the general population. Modern humans with their chain-link fences and wire cages find that isolation fairly easy to do, which is why most of the greatest number of breeds of dogs or breeds of anything are relatively recent.

Thus, artificial selection enabled the purposeful creation of most of our present breeds of dogs; they were created by crossbreeding types of dogs together in the late nineteenth and early twentieth centuries. Throughout the ages, sportsmen in particular have established and enjoyed dogs—often greyhounds or terriers or retrievers. They were not breeds as we know them; rather they were just types. Most breeding of these dogs was achieved locally for such basic reasons as the roads were bad and transportation to faraway places was difficult. The development of a breed was never thought of as making a purebred dog but, rather, of allowing the best working or sporting types to breed together, and then you culled what you didn't like.

Indeed, a common belief about dogs is that they are the most varied mammalian species. However, that is only true of the dogs in the show ring, in the AKC breed books, the pet play parks, and the pet stores. The most varied population in the world—ranging from two to two hundred pounds, with funny faces and funny legs and wrinkles on their skin or so much hair that it is hard to tell if there is a dog in there, and endless varieties of coat colors—are all mostly the results of crossbreeding. Crossbreed two different types of dogs and you don't get an average between the parents but rather you get something novel. At that point, one backcrosses the pups to a close relative or goes directly to inbreeding and sexually isolating the offspring for the production of "purebreds." These activities have nothing to do with the evolution of the dog.

The other way to produce a novel breed is to collect single gene mutations. When these mutations appear in wild wolves or coyotes, they are extinguished because the resulting animals are not the adaptive shape. Wolves born with the leg length of a basset hound are going to be selected against. It is only when humans support these mutations that a breed can be created.

Numbering only about 15 percent of the world's population of dogs, the purebreds, if left to mingle with a free-ranging population, are not going to have any effect on the genetic structure of the world population of dogs. If you released those 150 million purebred dogs into the wild all at once, they would not survive for very long. Why? They aren't adapted to any niche, and their numbers would be rapidly swamped by the 850 million village or street dogs, either by not finding adequate food or by crossbreeding with the wild dogs.

Our beautiful village dogs are neither the result of artificial selection nor the result of mongrelization of stray pet dogs. The similarity in size and overall design of all the street dogs and village dogs should give you a clue—a clue about natural selection. Why is that worldwide population of city street dogs and village dogs so uniform? For a biologist, that kind of uniformity implies the process of natural selection. Their size and shape (and even color) indicate an adaptation to a niche. The village dog is not a blending of purebreds that was created by artificial selection. The village dog is the animal that evolved on its own, with no reproductive control by humans, and is adapted to the niche in which it makes its living.

The message of this chapter is, those look-a-like dogs, in the same way as the look-a-like pigeons, have evolved right there in their niche and are uniquely adapted to this niche. They are not escapees from irresponsible dog (or pigeon) owners. They are a natural species that lives close to humans, finds its own food, and mates perfectly well without human control.

4

What Is a Niche?

Niche Defined

The reason the earth can support a population of 850 million "village dogs" is because human beings have — inadvertently — provided them with a niche. This niche is an environment in which dogs have survived for centuries. If humans were to disappear, the village dog niche would disappear, and dogs would go extinct. Dogs could not take up residence in the wild because that niche is already taken — by wolves, coyotes, and jackals.

Our task here is to demonstrate that the dog is a species of mammal that is adapted to its own particular environments. The dog is a shape that has evolved to a new niche that was created when people switched from hunting and gathering to growing grain. The waste products of that activity created a food supply that supports village dogs. Were there dogs before the age of agriculture? Probably not, but if there were, they had adapted to a different niche.

The term "niche" was first used in 1917 by Joseph Grinnell as "a place in nature." The analogy was to a recess in a wall where a statue stood. The statue represented a species, and the niche was where it fed. Ten years later, Charles Elton used the term as a place in the environment where a species fed and avoided predators. We would now say that adaptation to a niche focuses on foraging, reproduction, and hazard avoidance.

Since 1927, some intellectual giants in ecology and mathematics have wrestled with not only how to define and describe a niche but how to measure it as well. The Yale University ecologist Evelyn Hutchinson defined a niche in the 1950s as an "n-dimensional hypervolume." When our mathematical wizard colleague Mike Sutherland tried to explain it to us he said, "Yup . . . just a little piece of real estate in an abstract ecological 'space' . . . or . . . think of a niche as a little cloud floating around in seven-dimensional space. If you make twelve different measurements it would be twelve-dimensional space."

Professor Hutchinson graphed the physical properties of a lake, which wasn't just a lake but an arrangement of temperature gradients within the lake, and he illustrated how some algae could function and grow within a small temperature range, whereas a different species of algae was adapted to a different temperature regime. In broader terms, those organisms specifically adapted to a specific zone within the lake could not survive in another zone in the same lake.

Many writers after Hutchinson tried to perfect the definition and the means of measuring "a niche." Hutchinson was followed in his thinking by his student Robert MacArthur, who actually measured organisms on little tiny islands. One thinks of MacArthur for the concept of biogeography, where biological organisms are limited to a piece of real estate.

MacArthur teamed up with Professor E. O. Wilson to think about the evolution of survival strategies. An example is the concept of r- and K-selective strategies. We will discuss these in detail in chapter 8, but essentially dogs, an r-selected species, produce a lot of pups without much investment in parental care, whereas the rest of the genus—wolves, coyotes, and jackals—are K-selected species that invest an enormous amount of parental care in their few offspring. The wild types of the genus *Canis* are unusual in the mammal world in that both the male and the female care for the offspring and for a long time.

The discussion of what a niche is goes on even now. In 1983, Richard Lewontin at Harvard argued that niches are not simply empty recesses waiting to be invaded by an evolving species but rather that the species is an active participant in shaping their niche. The studies of wolves introduced into Yellowstone now use terminologies such as "trophic cascades." This has to do with a reorganization of plant distribution and even the changing of the flow rates and direction of rivers after wolves were introduced. Such transformations support Lewontin's thesis that the species not only adapts to a niche but also actually shapes it differently in the process of occupying it.

We agree, but it doesn't affect our argument that with the coming of agriculture, perhaps 10,000 years ago, humans created a big organic mess, a "trophic cascade," of new niches into which any number of species started the evolutionary process of adapting.

In the ecological sense, a niche is a piece of real estate that provides resources required for a species' survival. Within those habitats are other habitats, microhabitats and minihabitats, all of which can be the special environment of a particular species. The bottom line about the theory of niche, which ecologists seem to agree on, is that only one species can inhabit a niche at one time.

We often think of a species that is limited to, say, "the tropical rain forest." Looking at the earth from outer space one would see the tropical rain forest as a particular patch of color. However, when you are inside the tropical rain forest, you find smaller and smaller patches within patches. Just as Hutchinson's lake has species limited to certain temperature regimes, in the rain forest you find mosquitoes that are adapted to specific ranges of relative humidity and are only active during that period of the day when that particular band of humidity is present. We are all aware that when you are down by the lake at sunset, that is when the swarms of mosquitoes arrive.

The honey-making part of a flower is tiny, but there are, in the world, millions of flowers. Hummingbirds and bees have evolved organs to extract the honey from that tiny recess. The hummingbird is a specialist with its distinctive bill adapted/evolved to extract nectar from the flower. In the process, it fertilizes the flower. The relationship between the two species is called a symbiotic mutualism, meaning that both organisms benefit from the interaction. Thus, part of the flower's niche is the hummingbird. When studying the niche of a particular species of tropical flower in the rain forest, it would be mandatory to measure the activity of the hummingbirds that spread the pollen from one flower to another. And if the hummingbird population were to plummet, the flowers wouldn't be fertilized and in theory couldn't survive.

At the same time that the specialized bill and wings of the hummingbird are adapted to getting its calories from flowers, that very same evolved specialty limits the animal's ability to feed. It can only get nourishment from flowers. How many hummingbirds can exist is strictly dependent on how many flowers there are. This is called the carrying capacity of the niche. Again, this is part of MacArthur's biogeography.

One flower doesn't produce enough honey to feed one hummingbird. If it did, then the hummingbird would not move from flower to flower, fertilizing the flowers as it searches for food. Indeed, one might speculate that the amount of honey a flower makes is an adaption to how many visits the hummingbird needs to maximize the fertilization of the flower population. On the other side of that coin, there has to be a large enough population of flowers making enough honey to support a reproductive population of hummingbirds. Otherwise the symbiotic mutualism between the two species would not work.

That concept will become important when we discuss dogs specializing on human waste, where the same principles operate. A large enough population of people has to discard enough garbage or edible refuse to support a smaller reproductive population of dogs.

Looking at the niche in another way, each one must contain a pile of sugar, that is, accessible energy. What evolution is all about is developing a shape that can acquire that pile of sugar from a special niche more efficiently than any other shape. If, over time, individuals in a population change shape such that they are better able to capture a pile of sugar in a unique place, we say they have adapted to that niche. The hummingbirds and flowers are an example of that kind of evolution.

The basics: all life is dependent on photosynthesis. Green plants turn carbon dioxide (CO_2) and water (H_2O), in the presence of sunlight, into sugar ($C_6H_{12}O_6$) and free oxygen (O_2). That is the secret of all life on the planet. Sugar is the coal, oil, and gasoline of the biological world. It is the fuel that runs all biological systems. It is calories, which is another way of saying fuel. At our house, we have hummingbird feeders that are filled with four parts water to one part sugar. The feeders mimic the shape of a flower and provide the "nectar" hummingbirds need for their energy. We are a popular place for hummingbirds to hang out and fight over the available sugar. (Hummingbirds zoom about in airplane-like dogfights, defending the feeder, and we wonder if that's a good use of their energy.)

The basics for any species is to be a green plant in the sunshine and make your own sugar, or be an organism that is adapted to extract some of the sugar from such a plant, or be an organism that eats those that can get the sugar from the plant. Our ecologist friends call that a food chain. If any single organism doesn't get enough sugar, it starves to death — and a fact of life is that most animals do starve to death.

That, you may say, is oversimplified. Nope, not really. But what about protein, carbohydrates, fats, vitamins, and two green and one yellow vegetable per day? Different organisms make all those molecules, but they require energy to arrange those molecules into proteins, fats, you-name-it shapes. Sugar is the required energy.

In order to do anything such as reproduce or run away from danger, a living creature has to have the energy to do it — and the energy starts with plant sugar. If you want bacteria to grow on a petri dish in the lab, you have to put sugar in there. Moose eat green plants directly to get their sugar. Wolves eat moose to get the products of plant sugar such as fat. Moose have evolved lips and feet to get at particular kinds of plants that are hard to reach unless you have the lips and feet of a moose. Wolves have a shape to kill moose and get at the sugar (calories). The green plant, via the moose, provides the sugar for the wolf. Unsurprisingly wolves developed a digestive system by which they can convert fat moose meat back into sugar (energy).

Why don't other animals kill and eat moose and share the niche with wolves? Because they don't have a shape that is better than wolves to kill a moose in the north woods. Wolves have evolved the best size and shape for finding, killing, and eating moose and turning moose into wolf puppies. All the species in the animal world are competitive for a limited number of calories per sugar distributed unevenly over the surface of the earth in hard-to-reach recesses like the bottoms of flowers — each of which is called a niche.

The Niche Is Like a Fish Bowl

Let's look at this another way. The niche has a carrying capacity. What is true for hummingbirds is true for dogs. Picture a really large glass fish bowl full of marbles. The bowl is the niche and the marbles in the bowl represent one species adapted to that niche. We are thinking dogs. The marbles/dogs can feed and reproduce in the bowl but they can't live outside the bowl. The glass bowl represents a piece of ecological real estate, a combination of hundreds of environmental factors like temperature and climate and food and water and appropriate nesting places, et cetera, et cetera, et cetera: their niche.

Our imaginary bowl is full of blue marbles, and each marble represents a dog in a village. No two marbles look exactly alike but they are really close. They are about the same size, and none of them are perfectly round. They are all kind of blue, with maybe a little swirly white, but you would call them blue because they look bluish. Occasionally among this azure population appear a few odd-colored marbles that are almost all white or so dark blue they almost look black.

Now each marble in the bowl can breed with any other marble in the bowl, but each marble is most likely to breed with the marble next to it simply because, although it is not impossible to travel among the marbles inside the bowl, it is a tough journey requiring more energy. A marble could breed with a marble in another similar bowl if it could get out of its bowl and travel through an impossible landscape to another bowl in the matrix. All the marbles want to breed as often as they can because it satisfies a biological instinct.

If the bowl is full, there is no room for any more marbles. The reason for calling this niche a bowl is because it is important to understand the limits of how many marbles can fit in the niche. For every new marble that is born and grows up, a marble has to die or migrate out of the bowl. Or the one just born dies quickly because there is not enough space in the bowl for one more marble. The population in the bowl can never grow larger, simply because there is no more room in that niche. That is the point of representing the niche as a glass, and hence unstretchable, bowl. It cannot get any fuller. The bowl has reached its carrying capacity.

The niche can support just so many animals. The marbles continue to reproduce little marbles, most of which die. The only little marbles that can possibly survive are the ones that replace adult marbles that have died.

Then, one day, the population suffers the dreaded marbles disease, and the marbles die rapidly in large numbers. A lethal marbles disease goes through and kills thousands of them. They get marble mange or marble parvovirus or the deadly rabies. The disease makes the marbles weak, and so they have trouble capturing food and get weaker still.

Most often such diseases don't kill everybody, but they could, conceivably, kill every last marble in the bowl. The niche would still be there but empty. In real life, a reduction in population numbers is reasonably frequent and sometimes cyclical. Animals die off because of pestilence and natural calamities—a soaking wet year or a summertime drought. At this writing, wolves in Yellowstone are at their lowest population level since their reintroduction to the park in the mid-1990s. People are also watching the wolves on Isle Royale, an island in Lake Superior, where the wolves' numbers are dwindling to near extinction. Think of it this way: Isle Royale is a bowl of wolves, and scientists have kept records for decades on how many wolves inhabit the island each year. Populations commonly fluctuate, but in this case the number is declining, now down to almost zero. The question then becomes, should we reintroduce wolves into that bowl?

Our bowl will fill back up reasonably quickly. How quickly depends on how many have survived and their breeding rate. The latter is therefore important in understanding the differences in dog, wolf, coyote, and jackal populations. Dogs have a higher reproductive rate than the other species. The dogs can refill the bowl faster than their relatives can fill their bowls—but more on that later.

The bowl can fill back up in two different ways: first, all species over-reproduce, and when the niche is not full, over-reproduction is a good strategy. More of the marble offspring will survive than usual because of less competition for the available resources. Second, marbles migrate in from nearby bowls. Perhaps neighboring bowls are full and young animals are dispersing to find suitable habitats.

The trouble with analogies is they often sound stupid, but we will see examples in the wild types. In our studies, we found young coyotes sometimes dispersing hundreds of miles looking for an empty territory. Often they would be killed in the process. Why not stay and breed where they grew up? Most likely because the coyote bowl at home is full.

Before the die-off, the bowl was full of marbles. Every marble was slightly different in color, size, and shape. In fact, since all the marbles in the bowl are breeding together and are adapted to that particular bowl, the color differences between them might be really slight. After the plague, the new population might look almost like the previous one.

Yet more often, when we see this phenomenon in dogs, the new population will be colored somewhat differently. Let's imagine that when the population of marbles died off all that remained was one last atypical all-white pregnant marble. From that one pregnant marble and her descendants, the growing population will refill the bowl. When she got pregnant before the die-off she most likely bred with an average bluish marble, so the new population would still be bluish, but maybe more whitish than bluish. In other words the average color of the marbles in the bowl has changed.

Our white female is what is known as the "founder" of the new population. Because she was white, the new population tends to have a higher frequency of white dogs in it. Evolutionary biologists call this phenomenon the "founder effect."

Similarly, if every last marble died and a pregnant greenish immigrant came in from a nearby bowl, then the new resulting population would look more greenish. Again, this is also the work of a founder effect. Changes in characteristics are most dramatic when there is no real selective advantage for it. The color of village dogs does not seem to be very important; color patterns can vary a lot. In current populations of wolves or coyotes, the color we see has been selected for in the wild, and the animals tend to be uniform for very long times. In contrast, when we returned to Pemba years after our first visit, we were greeted by a whole new color scheme of dogs, but they were still the same size. That in itself indicates that size is an important adaptive characteristic and that color probably is not (but more on that later, too).

The founder effect "always" results in less genetic variability after a population decline. Founder effects are important for dog people to understand. Take a couple of simple examples. Someone starts a new breed by taking a few lovely dogs from a bowl. Let us say that when the collectors looked at the dogs in the Pyrenees Mountains bowl they noted some really nice-looking all-white dogs. They didn't collect dogs at random across the Pyrenees but sorted through them, only choosing those all-white dogs. They took the dogs to the United States and sexually isolated them from all the rest of the dogs in the whole world and called the results a breed.

Those few dogs will be the founding population of a new breed. They do not and cannot represent genetic diversity of the entire population of dogs they came from in the Pyrenees Mountains. Breeders tend to make this mistake, thinking that the AKC registered Siberian huskies genetically represent an ancient strain of dogs in Siberia. It isn't possible that the few dogs at Chinook Kennels in 1935 could genetically represent the whole population of dogs of Siberia.

A few years ago we collected Anatolian shepherd dogs in Turkey for our friends in California, who were starting a new breed to register with the AKC. It is a great story. We were going to Turkey to collect livestock-guarding dogs for our project. The question was, could we collect some Anatolian shepherds for them? "Sure," we said, "what do they look like?" They described them as big, tan dogs with black muzzles. Locally they are called karabash, meaning black head.

We sorted through the dogs in a vast area of Turkey looking for those dogs in a certain size class with that particular color pattern. Most important for our project, we collected dogs from working stock with the correct behavior to be livestock guardians. The two dozen dogs we brought back were a large part of the founding population of the new AKC Anatolian shepherd dog breed as well as potential livestock guardians.

But morphologically and genetically, those few dogs could not represent the gene frequency of the thousands of dogs in Anatolia. At about the same time, a U.S. breeder was collecting white shepherd dogs in Turkey, which he named akbash (white head). His claim was that, since many of the large livestock-guarding dogs in different countries were white, then white must have been the original color. Accordingly, he brought only the white dogs from the villages of Turkey to the United States and started a breed. Now that is a founder effect. It is the empty bowl of marbles where a few pregnant pure white marbles are placed in it. Then the claim is that these all-white dogs embody an ancient breed of sheep-guarding dogs, which you can tell because they are all white. Neither the Anatolian shepherds nor the akbash (and now rises another new breed, the Kangal shepherd dog) could possibly or even collectively represent the genetics of all the sheepdogs in Turkey—not now, nor from any "ancient lineage."

As we moved from region to region across Mediterranean countries looking for sheepdogs, we would see regional variations. In some regions, most dogs had rough coats and floppy ears and/or would be mostly white. In the next region, the coats would be darker, rougher, longish, and the ears erect. Why the differences in each region? It could simply be founder effects.

The founder effect turns out to be a breeder/hobbyist delight. Dog fanciers would and do go to a new region and sort through the different sizes, shapes, and colors of dogs. Our experience in Turkey resulted in white dogs being picked from one region and the tan ones from another. These dogs served as founding stock for new breeds in the United States and the United Kingdom. Then breeders would develop stories about this "ancient" and "superior" special breed, ignoring the fact that in these rural ruminant-raising regions nobody ever had reproductive control over their dogs.

Many of the differences in these landrace dogs can be attributed to natural selection for local conditions and/or founder effects. Another important process occurs. Shepherds often cull a litter to reduce strain on the nursing mother and to enhance the chances of healthy survival for a pup or pups that the shepherd favors. The couple of survivors of the culling might have a distinctive or regionally preferred coat or color. In these cases, coat color may be important.

For the moment, the important point is that the population in the glass bowl and all the other bowls can increase and decrease sometimes dramatically, but a maximum number of marbles cannot be exceeded in any bowl. Instead of counting dogs in the bowl, another way to gauge the population would be to measure how many pounds of dogs were in the bowl. Every pound of dog requires a given number of calories to grow from zero to one year of age, and every pound of dog has a maintenance cost. If, over time, the size of the average marble/ dog decreased, then there could be more marbles in the bowl as long as they were the same number of pounds. And, of course, if the size of the average dog increased, then the total number of dogs in the bowl would have to decrease, but the number of pounds would still be the same.

The bowl of marbles is one way of illustrating the Darwin principle—the number of animals born will exceed the photosynthetic production needed to feed them all. The ecologists put it another way, saying, "The niche has a carrying capacity." What they mean is: only so many pounds of dog/marbles can be sustained in the niche/bowl.

The bowl of marbles is a good way of conceptualizing the niche and its carrying capacity. But the niche is much more than a bowl crowded with individuals of one species. Within the bowl, a species' niche can vary in the quantity and quality of the required resources.

Think of a niche another way. It can look like an island surrounded by water (fig. 5, *top*). Our dogs couldn't feed and reproduce out in the ocean. The sandy beaches would provide only marginal sources of food, with perhaps a few dead fish and some crablike animals from which dogs could obtain a smidgen of nourishment. In the center of our island niche is a lot more food and possibly even a few hot spots that are really good.

The Mexico City dump is one of those hot spots for dogs (and pigeons). In 2012, the city ranked tenth in the world for population (nearly 20 million people). You can find dogs all over the city and its suburbs, but the highest density of dogs (that is, dogs per square kilometer) is at the dump.

Luigi Boitani and his colleagues, studying dogs in Roca di Mezzo, drew a map of the location of a small group of dogs and recorded where they spent most of their time during a three-year period. The diagram almost looks like an island even though it is right in the middle of Italy.

The dogs had areas for feeding, resting, and nesting. Rarely did you find them outside of this island paradise. Ninety-five percent of

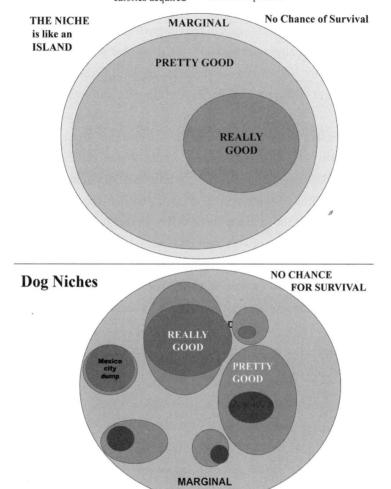

Figure 5. The niche is like an island. The species can't live off the island.
The resources within the niche are not necessarily evenly distributed as in the
top illustration but rather scattered throughout the niche as in the *bottom* illustration.
Even so, the species is still limited to the niche to which it is adapted.

the time they were on their island's home range. Some individuals migrated off the island or died, and a few immigrants replaced them. The total population of dogs stayed just about the same size over the three-year study period. These village dumps provide a niche (a hot spot) for dogs—whether you call them free-ranging dogs or dump dogs or anything you want—and it is capable of supporting a population of seven to ten dogs, year after year. Not only can it support that many dogs, but it actually does. Kill off a few as the local people sometimes do, and the area fills right back up with overflow immigrants from other dumps or niches.

Niches, like islands, have well-defined boundaries for a resource the dogs need in order to survive and reproduce. Within the island boundary is a food supply. That food supply is measurable in units like calories and—yes—other essential nutrients that another organism produces and the dogs eat. The food supply has to be obtainable. Our island might have tons of wildflowers with their little pockets of sugar that might be enough calories for a small reproductive population of dogs—if only they could somehow capture it. Or maybe on our island are tons of little seeds—enough little hard-shelled, tiny seeds containing enough calories for a reproductive population of dogs. However, the shape of dogs' mouths isn't as good at picking up seeds as are pigeon bills. The pigeons would outcompete the dogs simply because they are more efficient at picking up that food.

Even if the dogs could eat the seeds, they probably couldn't digest them. Our Jack Russell terrier would sometimes invade our birdseed supply and his poop would contain all of those seeds unchanged, still ready for the birds to eat. Back when there were more horses than cars, the domestic house sparrow population was locally enormous since they could pick the undigested seeds out of horse poop.

This is a great example of a house sparrow niche. Horses range over a large area eating grass, or grass from a large area is brought to the horse. With that grass come thousands of horse-indigestible seeds, which the horse concentrates and deposits in piles of waste. Those undigested islands of seeds are a rich food source for domestic sparrows. Like hummingbirds searching in flowers for honey, the sparrows search the farmyards of horses for piles of seeds.

When the horse population declined at the beginning of the twentieth century with the tremendous increase of automobiles and mechanized farm equipment, the sparrow population declined. They were considered pests at high population levels, but once their food source decreased, they became more manageable.

If you assume that an average thirty-pound, moderately active dog needs a thousand calories per day, then you can measure the number of calories in the dog food supply, divide by a thousand, and come up with the total number of dogs the island can support. If the food supply is human waste, then how much human waste accumulates depends on the number of people and how rich or poor they are. The illustration is identical to the house sparrow example. Humans collect grain over a large area, and even after cooking not all is digestible, and some of it spoils and with other food wastes is disposed of. That collection and disposal becomes the food source for dogs.

How many humans creating waste are enough to support one dog? As noted earlier, in some African villages we recorded seven dogs per hundred people. All other things being equal (which they never are), that implies that it takes the waste of fourteen people to feed one dog in that village. If our island had fourteen people living on it, then they would/could support one dog. But one dog isn't enough to constitute a breeding pair. Therefore, the niche is not big enough to support a population of dogs.

If the Mexico City dump can support 700 healthy dogs, then 700,000 calories must be delivered there every day. If Mexico City was like one of our common African villages it would take 10,000 people to produce that waste. It doesn't take long to figure out that Mexico City has many dumps. At the last count, there were 1,200 illegal dumps in the city. If all other things were equal then the 20 million–plus people could provide enough food to support 1.5 million dogs.

If the Mexico City garbage collectors go on strike, then dogs are im-mediately in trouble. The size of the food supply in the hot spots is dra-matically reduced in only a matter of days. This kind of event happens in almost all animal populations where seasonal bottlenecks result in food and water becoming unavailable for a period of time. Either ani-mals have to migrate to new food sources, the way sheep move from summer grazing areas to winter pastures, or they starve. Believe it or not, dogs starve very well. If they are still, they can survive a long time without food. If they were like wolves and searched far and wide look-ing for food and didn't find it, they wouldn't last as long. One of the neat things about a dog is they don't tend to search far and wide for food. The tendency is to just wait for the food to come to them.

Sometimes the hot spots are not continuous with each other, so the density of animals that occupy an area will be uneven over the entire niche (fig. 5, *bottom*). Some zones are more productive than others, meaning that a few areas of the island niche produce more dog food than other areas. Out behind the abattoirs is a good place to find high-quality dog food, maybe as good as the dump. In good and in really good areas, you would expect to find more dogs. As you head for the margins, you find fewer dogs, until you get to the edge of our island niche — and no dogs live out there in the water.

Have we exaggerated the argument even a little bit? Not really. Of course, the one hundred dogs on our hypothetical island need a little more than straight calories (sugar). They need a little protein (some essential amino acids, meaning those amino acids they can't make themselves). They also need essential nutrients and vitamins. If dogs are living on human refuse, none of those other essentials are really a worry. As human refuse rots (we call it dog yogurt), the microorgan-isms produce a good crop of amino acids (protein) and essential vita-mins.

One could postulate an environment where people could live but dogs could not. A hundred dogs need 100,000 calories a day, and they also need water. With a group of students, we studied islands called coral rags in the Indian Ocean near Tanzania. Coral rags consist of rock that is formed from porous coral. Rain water goes through the rock as if through a sieve. Rain water on the island settles down onto the underlying salt water, displacing it and causing a lens of fresh water. Plants with a taproot system can suck up some water, especially where topsoil has accumulated. Soil is a great reservoir for water—for plants, but not for dogs or people. Some plants are able to estivate or live on stored water during the dry season when it is not raining. But people and dogs can't estivate and they have to find water practically every day.

The couple of hundred people living on one of these coral rags can collect rain water from the roofs of their house and often they have large cisterns to store water for many days. It works well in the rainy season, but dry seasons lasting as much as half a year are tough. During much of the year people bring water to the island with motorboats. Water is precious, and a person might not even share with another person unless it was her own child or close relative.

On these coral rag islands people can grow some grains or starchy crops like cassava, but there is little to no edible protein produced on the island. Their protein comes from fish from the sea. Fishing is the job mostly of young men, or in biological terms, the job of reproductive males. Since they are the ones that catch fish, they have first access to that fish. They tend to get what protein they need.

Fish are valuable and are sold directly off the boat for cash. Merchants from Mombasa, Kenya, will tour with boats among the fishermen to see what they have for sale. The fishermen need the cash to buy gas for their outboard motors, which they need to bring water, among other things, to their families. These are people who live on the margin of our island niche.

And many people on these coral rags, in fact, show signs of protein deficiencies. Little kids and older people quite often have the gurgling cough of someone with congestive heart failure, as they digest their own bodies for protein. They tend to get enough calories but not enough protein. This isn't true of all of them. As we noted, the fishermen were in good shape, and, no great surprise, so were their girl-friends. The reproductive portion of the population was doing fine, but older people and children were in a bit of trouble.

The same story is true of dogs across their biogeography. The reproductive animals in an environment with scant resources tend to look great, but the juveniles most often look terrible. The adult dogs are the adaptive size and shape for their niche, but the juveniles have to compete with the adults for the essentials, and they always lose. More are born than the niche can support. Dogs are our topic, but if we were writing about any other species the story would be similar.

The people who live on this coral rag are basically a marginal human society. They occupy the margins of the human niche. Some reproduction occurs, but growth tends not to be very good, which leads to high mortality among the little kids. If our bowl of marbles represented people on a coral rag, then the bowl is full. So breeding and reproduction take place, but the little kids just die. As Darwin says — there is not enough food to feed them all.

Why are we talking about people and not dogs on these coral rags? Well, we saw no dogs on these islands. On one we studied, there were 200 people, which should have been enough to support fourteen dogs (a small reproductive population). But there is no standing water, and the people were not willing to share their water. A population of people, with no dogs at all is highly unusual.

The main Mexico City dump is a large landfill. Landfills don't have standing water. This is a problem for dogs and people in the Mexico City dump. Indeed, as on the coral rags, water must be trucked into the Mexico City dump for the 200 people who live there, plus the day workers. But water doesn't have to be shipped as far as it does on the islands of east Africa, and it doesn't cost as much. Also, the people at the dump are by far better off financially, and some of them share their imported water with dogs. The other advantage for the dump dogs is that their food is really fresh and juicy, delivered daily. They also can walk off their island niche to a local source of water. At night, many dogs follow dump workers home, where the dogs then get water.

In the Mexico City dump, dog food migrates into the dump daily. On our huge island called North America, food migrates from one area to another in very much the same way it migrates to the Mexico City dump.

Food for our dogs migrates to our house in the form of kibbles, from somewhere, perhaps the Kansas wheat or Iowa cornfields and Chicago slaughterhouses. Who knows? None of the ingredients of our dog's food comes from our yard or town or even our state. And we don't leave any garbage in our yard. Nothing around our house or our neighbor's houses would support a population of dogs—except the purpose-made food that comes in bags labeled dog food.

It is almost the same at the Mexico City dump. Seven hundred thousand calories of dog food arrive every day, and the dogs don't know where it came from or how it happened, nor do they care. The daily arrival there of that food is just part of their niche. Not all the food that arrives at the dump is edible by dogs. Some of it is too small—tiny pieces of food or the seeds from vegetables that dog teeth and tongue cannot ingest or digest easily. These bits are eaten by pigeons.

The Mexico City dump has more pigeons than dogs, and yet dogs and pigeons feed side by side barely noticing each other. Dogs and pigeons are eating from the same food source, but they specialize on a size range of that food. At one level, the dogs aren't competing for food with the pigeons. If the dogs disappeared, the pigeons might actually eat some of the food that dogs eat. When wolves disappeared from Yellowstone decades ago, coyotes would eat food normally eaten by wolves. When wolves were returned to Yellowstone, the coyotes went back to eating coyote-sized food. In the behavioral ecology trade, this is called resource partitioning. The pigeons are earning a living in a different niche in the same dump. They could eat some of the dog food, but they are not as efficient at it and it isn't worth arguing about. The dogs are not competing for food with pigeons. A dog's biggest competitor is other dogs. We are back to our bowl of marbles.

Dogs are a population of animals with their own distinct shape, living in their own distinct bowl. Biologists assume that an animal's shape is an adaptation, meaning that shape is the most efficient way to survive in their "n-dimensional" niche. The claim here is that our 850 million sexually reproducing population of look-alike dogs are a species-specific shape adapted to a niche. One doesn't have to assume any complicated stories about how dogs were domesticated. The evidence shows that they are a perfectly good ecological species adapted to a niche. They are their own species. They have been selected for their shape and size and color by nature. They have evolved for thousands of years, on their own, in a multidimensional hypervolume space that provides the scope and resources for this canid species to thrive.

PART II

Behavioral Ecology

5

Behavioral Ecology of Dogs

We have made the case that, sometime in the not too distant past, one or more of the wild types of canids evolved — evolved a dog shape and behavior. In this book, we are not worried about from what population that wild-type something evolved (wolves, coyotes, jackals, dingoes, or crosses thereof). Whichever it was, it was a population of animals of the genus *Canis*.

It is intriguing to try to understand what the ancestor of the dog might have been, but this question makes little or no difference to our discussion here. Many a book has been written on what the possible ancestor was and what the timing of that evolution might have been. As we said, these are affectionately known as the when, where, and how questions. They are entertaining questions, about which geneticists, paleontologists, archeologists, and mammologists, among others, all have their own evolving ideas: when, where, and how the transformation from whatever-it-was to dog took place. The theories of each group of scientists change from time to time, as new evidence and ways to measure that evidence are discovered.

We, however, are interested in something different. We are interested in why dogs evolved the shape and behavior we now call dog—without forgetting that evolution is an ongoing process. What we have done is to study living dogs around the world. Our focus is on what they look like and how they earn a living. What makes them so fit to live in such great numbers around the world? It is such a hard subject to write about for people in developed countries because the widespread conception of dog is usually of a recognized breed of dog, purchased from a breeder, which eats dog food purchased from the grocery store.

When one shape evolves into another shape, the Darwinian assumption is that the new shape is an adaptation to a new niche. A new niche means a new source of food and nesting places and a good location to raise pups, where the dangers aren't overwhelming.

The behavior of dogs (or any canids) can be studied in two basic ways: (1) by comparative psychologists who tend to study the *nurturing* of behavior, including development and learning, or (2) by ethologists who study the biology or *nature* of behavior. Within this latter group are behavioral ecologists.

One definition of "behavioral ecology" can be found at a web page recently created by the International Society for Behavioral Ecology: "Behavioral ecology is the study of the fitness consequences of behavior. Research in this field poses the basic question: what does an animal gain, in fitness terms, by doing this rather than that? It combines the study of animal behavior with evolutionary biology and population ecology, and more recently, physiology and molecular biology. Adaptation is the central unifying concept" (http://www.behavecol.com /pages/society/welcome.html; last update, August 21, 2013). For our purposes in this book, the word "fitness" in the preceding definition can be read as "survival."

Our approach here is to concentrate on how the animals feed, how they reproduce, and how they avoid hazards. The wolves in Yellowstone

National Park feed mainly on elk and bison, they den in caves at the beginning of April, and they take care of their pups for sometimes almost two years. The wolves on Isle Royale (the large island in Lake Superior) feed on moose. Both places are fascinating examples of how wolves make a living, and studies of these wolves have gone on for years.

For the behavioral ecologist it isn't really important if wolves eat elk or moose because they learned to feed on elk or moose or because they are genetically programmed to feed on elk or moose or both. The behavioral ecologist records how many elk wolves kill and what kind of elk—are they young, old, or sick? The behavioral ecologist measures the age structure of a wolf population, the survival rate of different age groups, and the kill rate on their prey species. They are interested in how wolves forage, how they reproduce themselves, and what kind of troubles they get into while doing all that. They want to know what wolves are feeding on and not why they feed on that.

Behavioral ecologists are like economists who discover what's going on by applying a basic formula. They solve for X: What is the benefit (B) of behaving (in calories), divided by the cost (C) of behaving (in calories), or simply $B/C = X$.

Don't go away—it is a very simple formula. The coyote chases a rabbit and burns X number of calories (sugar) doing it. That is the cost of behaving. The coyote catches the rabbit and eats it, getting calories back from the rabbit. That is the benefit of behaving.

Or, the coyote chases the rabbit, costing X number of calories. The rabbit escapes and the coyote obtains no benefit from its expenditure of energy ($B = 0$/cost of chasing).

Or, a big fat elk falls off the cliff right into the wolf den and dies. The wolves eat it. In this last scenario, the cost formula comes out: $C = 0$, and the benefit B is enough food/energy for two weeks for all the wolves and their pups.

The behavioral ecologists plus other evolutionary biologists assume that survival (of the fittest) depends on the benefit of behaving divided by the cost of behaving being greater than 1; in other words, $B/C > 1$.

Very simply, the calories-in produced by behaving are more than the calories-out expended by behaving. The individuals with the highest B/C ratio have a better chance of leaving more offspring than the animal with the lowest B/C ratio.

If the cost is greater than the benefit $(B/C < 1)$, then the animal is starving to death.

Putting it all together, as Darwin pointed out, more animals are born than there is food to feed them. In other words, for most animals, the cost of behaving (searching for, chasing, and killing prey) is going to be greater than the benefit of behaving (calories in the prey). For most animals $B/C < 1$, and therefore most animals are starving and won't reproduce—no pups. Darwin theorized that a few individuals behave so efficiently (they are better fit, or adapted), that the benefit will be greater than the cost. Those animals will survive and reproduce. The formula $B/C > 1$ is the simplest illustration of Darwin's rule of survival of the fittest, natural selection, and adaptation (fig. 5, *top* [chap. 4]).

Let us run this argument again. The dog evolved, meaning it changed shape from whatever the ancestor looked like. We are assuming that happened because the new shape was an adaptation to a new niche. Adapting to a new niche means the new shape was more efficient in that new niche than the old shape. The new dog shape was more likely to achieve $B/C > 1$ than the ancestral shape.

For evidence of that, we have argued here that dogs all look alike. Their new evolved shape and new behaviors are more efficient in this niche than their ancestral shape and behaviors were. We understand that the argument may be simple and our long explanation may be tedious. We also understand that shape efficiency is not the way very many people think about the 850 million dogs running around out there looking like mongrels and strays and products of irresponsible ownership. For the moment, assume that the street or village dog is a very sophisticated design.

Behavior (for the purposes of this book) is simply defined as an animal moving through time and space. It takes energy to move through time and space. How efficiently the animal does this depends on its design. A bird is designed to move through time and space by flying. It has a shape (design) that doesn't cost the bird very much (in calories) to fly through space and time.

That seems simple enough, but flying or walking or chasing is not the only cost. Add in an energy cost in building the organism from a single-celled egg to the reproductive adult; this is called the design cost. How much does it cost (in calories) to build an elephant or a bird or a dog? Is that a different cost (in calories) than building an earthworm? Of course it is. And it costs much more to build a hundred-pound wolf than it does to build a thirty-pound dog.

Thus the behavioral ecologist wants to measure four major costs for any animal or species: the costs of design, foraging, reproduction, and hazard avoidance.

6

The Cost of Building a Dog

Look at the picture of the dog and the wolf (plate 4). Many people look at the wolf and think, "Now that's a predator." It is beautifully adapted to hunt big animals like moose. The dog is typical of most of the dogs seen around the world. It weighs a bit less than thirty pounds. The wolf spends its summers on the tundra and its winters on the taiga at the border of the northern forests of Quebec. This one weighs about seventy pounds.

Obviously, between a dog and a wolf, the wolf is the more expensive animal to build. The wolf is two and a half times the size of the dog. How much energy does it take to build the dog? Maybe a thousand calories per day. If all these things were linear, which they usually aren't, it would take 2,500 calories a day to build a wolf. Maybe it takes less than 2,500 calories a day when the wolf is a pup, but not much less, because growth and the speed of growth are energetically expensive.

The wolf does not become a successful hunter or come into reproductive readiness until it is almost two years old. That means it doesn't
get to its adaptive adult shape until it is two years old. Think about the
two skulls in the first chapter (fig. 2, *bottom* [chap. 1]). If the skull on
the left is the perfect shape, then the skull on the right has almost a year
to go to have the selected shape.

Just for ease of thinking about it, assume that the design of the wolf
requires 2,500 calories a day to grow up. It needs less at the beginning
and more at the end of the growth period. The average wolf pup needs
2,500 calories per day until it is nearly two years old (365 days × 2 years
× 2,500 calories) = 1,825,000 calories. It takes close to 2 million calories to grow a wolf pup to the size and shape at which it could efficiently
kill its first caribou. That number is artificially low because we didn't
calculate the extra calories dad and mom (and sometimes older siblings) added to their daily diet to feed junior to full size. Part of the design cost is the calories mom burned up making the milk, and the number of calories mom and dad used up to catch calories, digest them into
a usable form, and then regurgitate those calories to the growing pups.

The design of the dog, in contrast, requires a thousand calories per
day (less when a tiny pup) for many fewer days to become an efficient
acquirer of food than does a wolf pup. A young dog can begin obtaining its own food as early as ten weeks of age. In the Mexico City dump,
pups start scavenging food shortly after they are weaned. We figure
1,000 calories × 70 days = 70,000 calories. The dog needs 3.5 percent
of the calories a wolf needs to get to the point of collecting their own
food. Think about that. If I drop a seventy-day-old wolf in the middle
of a caribou herd, what are its chances of surviving? Just about zero. If I
drop a seventy-day-old dog pup in the Mexico City dump, what are its
chances of finding enough to live on? Not great—but a lot better than
the wolf pup among large robust ruminants (fig. 6, *top*).

The wolf pup becomes sexually active in its second year, taking
2 million calories to get there. The dog can become pregnant at seven
months (after consuming 210,000 calories) or 10 percent of what

Figure 6. *Top,* A pup in the Mexico City dump trying to make it on its own and not doing very well. *Center,* Our vulture feeder, baited with roadkills, is a habitat the vultures can exploit though our short-legged dog cannot. (Photo by Tai Coppinger.) *Bottom,* In Ontario, a fox raided our campsite and made off with a carton of milk: a scavenger at heart.

wolves require just to get bred for the first time. Getting the dog to reproductive age costs a small fraction, in terms of calories, of what it takes to get the wolf to that first litter of pups.

In other words, the design of the dog means it is cheaper to build than that of the wolf.

Why do we compare the design cost of the dog with that of the wolf and not with the coyote or jackal, which are closer in size to the village dog? The enormous differences in the costs of wolves compared to dogs exaggerate the discussion in order to emphasize the way a behavioral ecologist thinks about the fitness of animals for their niche. But even if we had used the twenty-five-pound coyote or jackal to illustrate our point, we still would find dogs to be less expensive in comparison by the time they reach hunting and reproductive ages. In part, this is because if the adult shape is the perfectly adapted shape for hunting, then anything different from that shape is not as good and will be outcompeted, and therefore the juvenile is always at a disadvantage simply because it doesn't have the adaptive shape until it grows up. Consequently, the design of coyotes and jackals includes help in calories from mom, pop, and often an older sibling, which is not true of the dogs.

There are, however, a number of ways of looking at design cost. For example, different species of ungulates feed on specific types of vegetation that differ from species to species. Each species is a specialist and can do something the other species can't. An obvious example is the giraffe, which eats leaves that are high up in the trees. It isn't that giraffes couldn't eat grass but, rather, that they don't exactly have a shape that's especially efficient for eating grass. Just getting that head down to drink water looks like a chore.

Or contrast the giant panda bear and the little red panda. Both species live in similar habitats and both eat bamboo. What is remarkable about the two species is that each is adapted to eating different parts of the bamboo. They therefore have different niches. The little red panda eats only newly emerged tender sweet leaves. Its delicate mouth is just right for picking little leaves. The giant panda crunches the stalks of the older parts of the plant. Their big teeth grind up volumes of junk bamboo and swallow them.

The bamboo that the big panda eats is low-quality food and so the big guy has to eat loads of it to get enough nourishment. That raises the question, Why is the big panda big? Well, it has to be big to get enough of that low-quality food in order to ingest enough calories to enable it to move through time and space. But the young giant panda is at a distinct disadvantage simply because it requires a lot of calories for growth and is not big enough yet to consume a sufficient amount of a low-quality food.

Do all species of mammals have this problem? Though biologists "always" hate to say "all," the answer is yes.

So why is the little red panda so small? Well, because although new leaves are good-quality food, not so many leaves grow in abundance, and the job of picking them is laborious.

Lorna's father used to say, "With patience and fortitude the lion bit in to the mouse," hoping his offspring would get the pun intended in the word "to." With comprehension would come the realization that a 300-pound lion jumping through the air to catch and eat four ounces of mouse is a waste of energy, as doing so would take more energy than the lion would get back from its meal ($B/C < 1$). Lions would starve to death eating mice just as each of the panda species would starve to death eating the other's diet. With patience and fortitude, the giant panda could slowly pick the small leaves, while the little red panda could barely get enough old stems chewed up to get an adequate diet.

Little red pandas have to compete for the leaves with other little reds, and all the same rules are in place: $B/C > 1$. The little panda has this amazing calorie-conserving design, which lowers its energetic cost. Their legs hibernate overnight when they sleep. The actual metabolic temperature of the legs will be ten degrees lower than the rest of its body. That saves energy that doesn't have to be replaced by eating more little green leaves. Wolves have a similar adaptation, where they can re-duce the blood supply to their legs and feet when it is cold or they are standing in cold water. That way they can keep their feet above freezing temperature without spending all those calories keeping their internal temperature up to 101.5 degrees Fahrenheit.

Evolution is about competition for a limited number of calories. More animals are born than there are calories to keep them all going. Every little adaptation helps. The fittest survive.

Stomach design and size is critical for species such as the canids. The average dog and the average wolf have a finite stomach size. Thus wolves and dogs can only eat and digest a limited volume of food per animal per day depending on the size of their stomachs. As with the pandas, the calorie content of that volume becomes a critical variable. If wolves fill their stomachs with mostly junk (for example, hide and hooves), then $B/C < 1$ even on a full stomach.

A wonderful illustration of energy in/energy out versus stomach size is dogs running in Alaska's Iditarod Trail Sled Dog Race. In 2014, the 1,100 miles was completed in eight days, fourteen hours, and nine minutes, which equals 130 miles per day, or 5.5 human marathons per day for 8.5 days. It is calculated that each dog can burn 10,800 calories a day accomplishing that task. The problem is that their fist-sized stomachs are not big enough to take in that many calories all at once no matter how good the food is. Thus, in order to be able to run that race, the driver must feed the dogs many times a day on foods such as fats and oils that are rich in calories.

It is also important to grind the food, so that the food has thousands of digestible surfaces. Grinding up the dog's food enables the food energy to be more quickly digested and thus room made for the next meal. Remember mom telling you to chew your food well? She knew that the little stomach of a rapidly growing child needs all the digestion help it can get. Left to their own devices, dogs, like wolves, tend to swallow big hunks of meat. But imagine a two-pound steak sitting in a sled dog's stomach for a day and a half while the dog is running.

A really keen sled-dog driver would warm the food and water to body temperature. That way the animal wouldn't spend precious calories heating its food to its body temperature, and digestion could take place more efficiently. If you feed a dog a quart of ice, it has to spend 160 calories heating that ice up to 101.5 degrees Fahrenheit. That's 10 percent of its resting caloric needs for a day in the arctic cold.

Energetically, it is not worth feeding a quart of ice to a dog, even though the dog needs the water. Quite often, hard-working sled dogs grabbing at snow for water will show the signs of dehydration. The same must be true for wolves in the frozen north, except they get most of their water from the flesh of their kills. Eat it while it's hot! Imagine trying to run the Iditarod race on dry dog food and cold water.

Most wolves starve to death. The 850 million village dogs' pups starve to death. Most individuals of most wild species starve to death. All species, as Darwin wrote, over-reproduce. Individuals of a species are competing with other individuals of their species for the same food (same niche). The niche has a carrying capacity. Thus when the niche is full and individuals in a species are over-reproducing, they will all be skinny and susceptible to starvation-related problems such as diseases. It is the bowl of marbles again.

The point here is that the design cost (in calories) to get a dog up to the right shape (a design) so that the dog can feed by itself is enormously different from that of the wild types. Dogs and wild canids are adapted to different niches and eat a different food. Just like the giant and red pandas, dogs and wolves occupy different niches and are designed to live and feed in their own niche. The red panda eating big old bamboo leaves and stalks would quickly starve to death no matter how much it ate since it doesn't have enough stomach size to digest big old leaves. Similarly, the dog would starve to death trying to hunt caribou for a living.

One of the windows in our home looks out over a field where deer feed. Quite often in December we'll say, "They don't look good. There are too many of them and the population should be thinned—are they going to make it through the winter?" Even if they all make it through the winter, will they have enough energy to make milk for Bambi?

What does that mean for the Bambi crop in the spring, and is the population going to crash because they have overeaten their food supply? A few years ago the George River caribou herd (in northern Quebec and Labrador) did just that—it overate its winter feeding area. When the herd returned in the fall, the lichens had not recovered, and thousands of caribou starved to death.

We recognize wolves, coyotes, and jackals as singular species, and we see that they are adapted to different niches. The reason they are different sizes and shapes is because they are adapted to (meaning evolved to or designed for) those different niches. The same is true for the dogs. They are a beautifully designed species and have a design that is more efficient for their niche than is that of their canid cousins. We'll talk more about village dogs outcompeting the wild ones at the dump in the next chapter.

Plate 1. Look-a-like dogs scavenge food, water, shade, and other essentials from humans. That is their niche the world over. *Top*, Benin. (Photo by Alain Weiss.) *Bottom*, Mali. (Photo by Kristina van Haagen.)

Plate 4. The wolf in northern Quebec (*top*) and the dog in Namibia (*bottom*) have different foraging and reproductive strategies.

Plate 2. Village dogs around the world all tend to look alike. This has to be the result of natural selection. *Top left*, Mali. (Photo by Kristina van Haagen.) *Center left*, South Africa. (Photo by Daniel Stewart.) *Bottom left*, Cuba. (Photo by Jane Brackman.) *Bottom right*, Vietnam. (Photo by Alain Weiss.)

Plate 3. Feral dogs playing tug-of-war in the Mexico City dump. For those dogs that get to adult size, living in the dump is full of fun.

Plate 5. Dogs don't go looking for food the way wolves do. Pictured here are dogs in the Mexico City dump waiting for food to come to them.

Plate 6. A great deal of food arrives at the dump daily and is always placed somewhere new. Thus there is little cause for fighting. (Photos by Virginia Dare.)

Plate 8. Two dogs in a Zulu village. Are they pets? Are they neighborhood dogs or one of each or both sometimes? (Photos by Daniel Stewart.)

Plate 7. Endless social behavior in paradise. Wherever village dogs lead rich social lives, it's mostly about reproduction, and they don't tend to fight about it. *Center*, South Africa. (Photo by Daniel Stewart.) *Top* and *bottom*, Mexico City dump.

Plate 9. A veterinarian arrives at a South African township and tries to vaccinate all the dogs. (Photo by Daniel Stewart.)

Plate 10. This Zulu lady has been adopted by the dog. (Photo by Daniel Stewart.) *Inset*, In the Mexico City dump, some workers are adopted by many dogs on a daily basis. (Photo by Virginia Dare.)

Plate 11. *Top,* Dogs adopt people in the dump—might be different dogs or different people every day. (Photo by Virginia Dare.) *Bottom,* A Rottweiler in the dump might be a stray, or might be a day migrant, or might be owned by someone living in the dump. Whichever, it can father some dump puppies.

Plate 12. *Top*, The favorite puppy in a Lesotho village has a better chance of survival than the not-so-favored puppy, looking on hopefully. (Photo by Tim Coppinger.) *Center*, Children have the time to play with animals; these Tanzanian children have adopted a nestling bird. The problem is that little birds rarely survive adoption. *Bottom*, Many species are adoptable and trainable like this pig, which is being taken for a walk in Mexico City.

Plate 13. *Top*, The lady lives in the dump and has purebred dogs that she cares for. *Bottom*, This terrier was a permanent resident and did not appear attached to anyone in particular—a real stray dog. (Photos by Virginia Dare.)

Plate 14. Hybrids. *Top*, Brother and sister coyote × beagle at Hampshire College. (Photo by Frank Mantlik.) *Center*, Jackals and dogs are sometimes hybridized for use by customs officials because they have a better sense of smell. *Bottom*, Hybrid wolf-dogs are the results of a German shepherd being bred to an Italian wolf at the zoo.

Plate 15. Many species of wild animals—from mice to elephants to dolphins—
can be tamed and trained. *Top*, Monty Sloan of Wolf Park plays with a wolf.
Bottom, World-famous animal trainer Ken McCort says that wolves learn more
quickly and remember the lessons longer than do dogs. (Photo by Monty Sloan.)

Plate 16. Wild turkeys being fed by hand. Domestication is not a great mystery. Taming and training crop pests and scavengers is easy. In our yard the squirrels always seem smarter at solving puzzles than our dog.

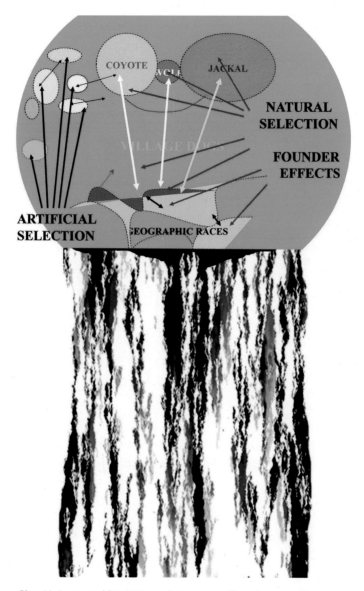

Plate 16. An atypical Darwinian evolutionary tree. The colors streaming up the trunk (through time) are populations of wolves, coyotes, jackals, and dingoes flourishing at times and going extinct or evolving into something new. At the top of the trunk, a massive population of dogs evolve (*black*), covering the whole world. At the top is the living canopy of mostly dogs with just a few of the wild types and a few breeds of dogs scattered throughout. (Design by Lee Spector.)

7

The Cost of Feeding a Dog

A big calorific cost for any species is finding food. In fact, the evolved shape of any species is almost always a signal about how it feeds and what it feeds on. The size of the animal, the shape of its mouth, and the shape and number of teeth are all clues for how it gets its food. Let us look again at why the 850 million village dogs all look alike.

As we suggested earlier, behavioral ecologists are really economists. They ask the one simple question, "What is the cost of behaving compared to the returns on that investment of energy?" Let's get a better sense of this using a simple example.

A big man runs a mile uphill on a really cold day. Let's guess it costs 250 extra calories to get to the top. This is what is called a transportation cost. At the top of the hill is a cup-sized jar of olive oil worth 1,700 calories. That replaces the 250 calories used to get there. He uses the other 1,450 calories for basic metabolic energy to run his body for twenty-four hours. He needs to store some of the calories as liver sugars and muscle fat. An animal, like a human, can use that sugar and fat for making the next trip up the hill tomorrow or for escaping from some dangerous predator, and he needs to save some for making babies later on. That seems simple enough. If just hanging around on a cold day costs 1,450 calories, the big runner is just breaking even.

Next day the man runs up the hill, and at the top is this tough little kid hanging onto the jar of olive oil; it costs 200 calories to wrestle it away from him. That is an acquisition cost. Drink it all (1,700 calories), and the benefit of behaving divided by the cost of behaving is $B/C = 1,700 \div 450 = 3.9$ — which is greater than one. Again, if we add the basic metabolism design cost, we are beginning to lose. It is getting to the point where it might be better to stay home.

In the same scenario, the guy runs up the hill on a cold day and the little kid sees him coming, kicks him in the shins, and runs away, and the man can't catch him because his leg hurts. Now it costs 250 calories (transportation) plus 200 calories (acquisition) plus the repair cost for the leg plus the basic metabolic cost. The benefit of moving up the hill is zero. The costs for the day are better than 2,000 calories. The person is starving.

If the runner were a little smaller it wouldn't have cost so much to grow him, and he wouldn't use so many calories transporting himself to the olive oil. And if he could sweet-talk the little kid out of the olive oil and the little kid didn't kick him, then the acquisition and repair costs would be minimal.

In the latter scenario, the transportation cost is less than in the earlier scenarios, the acquisition and repair costs are really low, and because the man is smaller, the design plus maintenance are less $(B/C > 1)$. Our sweet-talking little runner would outcompete the big tough guy. The silver-tongued lighter runner will have more energy to put into children than the heavy guy. The children in the next generation will be smaller than average and sweeter natured and will be able to collect lots of olive oil. Evolution in action.

It is a silly example, but this is essentially what the wolf, coyote, and jackal are faced with every day of their lives.

For a wolf or a dog, the cost of the transportation to food is measured in calories, plus the cost of acquiring that food (the killing cost for the wolf) measured in calories, and the repair cost measured in calories, divided into the benefit of the food acquired measured in calories, and the result has to be greater than one. The animal is losing its life unless $B/C > 1$.

The animal has to find and capture food. The food, meanwhile, does all kinds of interesting things. Food may be distributed evenly throughout the niche or it may be concentrated in some hot spot. Sometimes it is evenly distributed over a large area such as grasslands. Some species' food moves around. Often that movement is seasonal. Whales follow food sources as those organisms bloom in different seasons around the world's oceans. Caribou move from the boreal forest and taiga after the winter and spend the summer in the tundra to feed on lichens that emerge after the winter freeze. In northern Quebec, one of us studied wolves in an area where there were seventy frost-free days a year. That means there are only seventy days when photosynthesis can take place. Guess why no big trees live there? Wolves feeding on caribou have to follow the caribou every day of every season, which means a large transportation cost. Take, for instance, a pregnant wolf. She has to carry pups around as she hunts, adding to her transportation cost. And a yearling wolf is not very good at killing caribou because it is still working on fulfilling its design cost. It has to follow parents who are following food. Still, B/C has to be greater than one, even for the youngster.

Or, as mentioned, food might be evenly distributed. Picture a large population of mice distributed across an enormous grassland. A large population of coyotes making their livings eating these mice might also be evenly distributed across the same grassland.

Coyotes compete with one another for those mice. But, says Darwin, too many coyotes are competing for not enough mice. A male and female coyote and their young would have to defend a territory that has enough mice to cover their design costs, transportation costs, and any other costs. Each pair and their young would need to defend an area big enough to feed them all—that is, a feeding territory.

If the population of mice is low, then they would need a bigger feeding territory and that would increase the transportation cost to the farthest-away mouse. Big territories cost more to defend. Trying to enlarge a territory may be tricky because their neighbors may also be enlarging theirs.

With too few mice to support that lifestyle, then the young coyotes would have to disperse. J. Lorenz and G. B. Will studied an area in Vermont where the young coyotes would disperse a couple of hundred miles (as the crow flies) from their home site, presumably looking for their own territory. The animals would often be found dead over in New York State, which raises the question of how they got over or around Lake Champlain.

If you were a biologist studying coyotes and wolves, you might conclude that coyotes are solitary or territorial feeders, and wolves are more social feeders. You might say "wolves pack." And you, the biologist, might be wrong. Such innate behavioral differences might not actually exist between wolves and coyotes. Rather, they are feeding on a different distribution of their food supply that causes them to behave differently. When you watch three or four wolves chasing a caribou, you might think you're seeing a pack of wolves operating in concert. But it's also possible that they might be four individual wolves chasing the same sick caribou.

And if they catch it will they share it? The same rules about costs apply: without enough calories to fill up their individual stomachs, it would be cheaper for each wolf to share than to try to defend the surplus. Besides, if it is winter the surplus will have frozen by the time the first stomachful is digested. The next stomachful isn't worth as much because it has to be warmed up to 101.5 degrees Fahrenheit in the stomach. Thawing out five pounds of caribou for a day's feed is not energetically cheap.

Foraging Cost for Four Interacting Species

A simple illustration of benefits/costs of foraging became clear to us in a 1980s study of puma predation on cattle and sheep ranches in New Mexico. Often we would find cows dying, or victims of predation, right next to the water holes. Our study showed that those 1,200-pound cows (thinking: design cost?) didn't have the energy (calories) to move far enough from the water hole to find enough grass to provide enough nourishment to make it back to the water hole. The transportation cost of finding feed was too high: $B/C < 1$.

Picture it: you have a water hole, and the cows eat all the grass around it. The next day they have to go farther to find grass, and the next day even farther than that, and then they still have to get back to the water hole for water. In this desert scrub country, there wasn't enough rain and photosynthesis to produce enough plant growth to pay the cow's transportation cost to the feed and back to the water hole.

Also, walking so far and searching for grass produced sore feet, which further reduced the cows' abilities to find grass. The repair cost (also measured in calories) was too high. Finally, they would stop trying to walk to grass and would starve to death by the water holes. And just as likely, they became easy prey for pumas that were then accused of killing cows and hunted by the ranchers and killed as the rancher defended his own "calories."

The pumas hung around the water hole waiting for a cow, which reduced *their* transportation costs. The cows were weak and that reduced the puma's acquisition cost and the repair cost.

Sheep eat the same grass as cows. Sheep at the same water hole survive fairly well, however. First of all they are a smaller size (approximately eighty pounds) so they don't need as much grass. They also have an interesting physiological characteristic. Every time a sheep digests a molecule of fat, it gets two molecules of water. The same is true for camels. The reason camels can go so long without drinking water is they are digesting fat in their hump for both metabolic energy and water.

Breeds of sheep in arid places, such as Turkey, are called fat-tailed sheep; their tails are a resource equivalent to a camel's fat hump. That allows them to go several days without drinking water. Thus they can collect grass at considerable distance from water holes, surviving in the desert scrub for days without going back to water.

This New Mexican niche consists of a particular distribution of water and a particular density of grass per unit area. Sheep can survive on that distribution of grass and water and cows cannot. Sheep have "evolved" a design for living in that desert scrub niche. Or, one could just say sheep are better adapted to living in that environment. (One might ask the quick question here: If people domesticated sheep why didn't they select for large sheep—1,200-pound sheep with tons of wool?)

Of course, it is more complicated than just the distribution of grass and water. Sheep don't have to come to the water hole daily and can range farther away in search of grass. Some of the energy of looking for and acquiring grass has to go into paying the added transportation and acquisition costs. The good news, however, is that they can move so far that it increases the transportation cost for the puma hunting for them, which is bad news for the puma. Good news for the puma is that the sheep are 10 percent the size of a cow, making the puma's acquisition and repair costs less. But then the total number of calories in a sheep is far less than the calories in a cow, and the puma has to hunt more sheep to survive, which further increases the transportation and acquisition costs. Such hunting also makes the ranchers angry, and they become a serious hazard for the puma to avoid—staying constantly alert is also costly in calories.

The problem for sheep is that, because they are smaller than cows, they are more vulnerable to predation by coyotes, which can travel huge distances more cheaply than the puma can. A healthy cow is not really threatened by coyotes because the acquisition and repair costs are too high for a twenty-five-pound coyote. A healthy adult sheep is just about the size limit for a coyote prey, although lambs are easier prey for them. Since mother sheep have trouble protecting lambs from coyotes, when you look at coyote diets you find more lambs than adults. Because lambs are attractive to coyotes, ranchers hunt coyotes as well as pumas.

On and on the observations go. Now the puma is in competition with coyotes for the same available resource/prey: sheep. The sheep are in competition with local deer, elk, and antelope for the same grass. If the wild population of ungulates is high, then less grass is available for sheep. Therefore, the population of sheep must be lower and must range farther. Ranchers often encourage hunters to hunt on their grazing land because hunter success reduces the numbers of wildlife. Seven rabbits eat as much grass as one sheep, and thus they are a favorite sport hunt. Parties in Texas might include a sport drive with a bunch of people in the back of a pickup truck driving around shooting jackrabbits. In Europe and England, the landed gentry encouraged sport hunters with their lurchers and other "long dogs" to course rabbits — perhaps basing the hunt at a different farm every week.

The rancher now adds a livestock-guarding dog to stay with the herd or flock, and it is so good at warning predators away that neither pumas nor coyotes can get easily to the sheep. This reduces the energy that the sheep use to escape predation. The sheep with the dog are better adapted to the landscape (the niche) than are the deer. The population of deer decreases, which means the population of sheep can increase as they have less competition for grass. Because of the dog, coyotes have to switch to eating rabbits and mice. Coyotes have to spend more transportation time and energy and they have a higher acquisition cost while patiently waiting to pounce on one of the little rodents or chasing rabbits. If the coyotes would just eat rabbits and mice, then the ranchers would like them because preying on rabbits would reduce the competition for grass for the sheep. The advice to ranchers is, don't cull any coyotes that aren't killing sheep because they are actually helping provide more grass for sheep.

The puma population decreases because the prey base gets smaller with the loss of deer, which were not protected by the livestock-guarding dogs. Thus, both puma and coyote populations decrease in their former magic islands.

The sheep story is illustrative of how competition and selection vary in different types of niches. Compare, for example, a farm in England versus the New Mexican desert: a well-managed lush farm in England has so much grass one might keep seven sheep and their lambs per acre per year, whereas in a New Mexico scrub desert environment, it takes fifty-five acres of land to produce enough forage for one sheep and her lamb. In England, farmers select for sheep that will have twin lambs, while in New Mexico ranchers select and manage ewes for producing but a single lamb. In the dry environment with the larger spacing between blades of grass, twins means the death of both lambs simply because the ewe can't get enough nourishment to keep herself alive and produce enough milk for two lambs. In the rare event of twins, the desert rancher would cull one at birth in order to try to save the other. Even then the ewe might be so undernourished, having invested energy carrying two lamb fetuses around fifty-five acres (big transportation cost), that she might not have enough resources left to nurse the one remaining lamb. That is the very simple mathematics of the behavioral ecologist: B/C must be greater than one, or the animal is dead. Survival of the fittest. Having only one lamb is an adaption for surviving on desert scrub.

Cost of Feeding a Wolf

An adult wolf often has to search for hours and sometimes for days to find its prey. (A quick note: searching by wolves is often done with the nose. They can smell prey over a mile away.) Caribou wander over large tracts of land, and moose are often out in the water feeding, both of which scenarios add to the wolf's energetic problems. Some observers report that a wolf feeding territory can range over hundreds of square miles. Trying to find prey in that kind of area is energetically expensive.

How many calories does a wolf expend to capture how many calories? Let's say that six wolves averaging eighty pounds each have a feeding territory of 350 square miles. Now just make a budget. The 480 pounds of wolves expend thirty calories per pound per day searching for something to kill and then killing it (transportation cost, killing cost). The six wolves expend 15,000 calories per day. That means the group has to obtain 15,000 calories per day just to break even. If a pound of moose meat yields 460 calories per pound and the average kill (baby moose) is 200 pounds of digestible material, then the wolves would need to kill once every six days (200 pounds times 460 calories per pound = 92,600 calories, divided by 15,000 calories per six wolves). Each wolf has to eat an average of five pounds of moose per day.

Now we are back to a design problem. Imagine each wolf eating thirty pounds of moose once every six days. That is an almost impossibly large amount, depending on the size of the wolves, especially their stomachs. What happens is they eat what they can and defend the rest until they are hungry again. We now have a new cost: the cost of defending the carcass until their stomachs can hold the remainder. If the carcass freezes, they have the additional energy problem of thawing it out in their stomachs.

Defending the carcass turns out to be another big energetic problem. Letting a pile of meat lie around for six days invites all kinds of competitors to try to filch it. For wolves, the problems are foxes, coyotes, and ravens, which scavenge a good portion of any kill. They dart in and out when the wolves aren't paying attention. Then may come the dreaded bears, which simply take the carcass away from the wolves. Think how easy it is for the bear: no chasing of caribou required—meaning low transportation cost—just a big pile of meat lying there, and those "little" wolves trying to defend it.

In a study of wolves on Isle Royale (a study spanning fifty-plus years), researchers reported wolves killing one moose per week. At last count, 1,050 moose and nine wolves lived on the island. It looks like the wolves, for some reason, are going extinct.

Wolves also need to have extra stored calories in case they expend energy looking for prey and don't find it or in case the prey is too big or too fast for them to capture and it gets away. It is the little kid with the olive oil story all over again. Wolves need enough reserve energy to try to do it again. It is essential for them to have stored energy sufficient to make up for these unrewarding expenditures of energy. And, of course, the stored energy requires extra pounds and extra energy to carry it around. When things begin to go badly for a pack of wolves, problems begin to escalate. If wolves get a little run-down and weak, the chances of a successful hunt begin to decrease. Also, they are more susceptible to illnesses or accidents. The fragile Isle Royale wolves suffered a canine parvovirus attack in the early 1980s, and in 2012, researchers reported three wolf carcasses floating at the bottom of an abandoned mine shaft. As we have said, most wolves (and other wild animals) starve to death.

Some studies have reported that wolves specialize on the very young or the old or sick animals as prey. The theory says that this is a good survival strategy simply because young, old, sick, or injured animals don't have to be chased so far or so long, and they are less difficult and dangerous to attack and kill. Wolves that specialize in the old and the sick are not investing as much energy on transportation and acquisition costs, nor are they taking as big a chance in getting hurt (which would increase the repair cost) as are those wolves that go after healthy prey.

Hunting the young, old, and sick has a down side, though. Yes, less energy is involved in the hunt and kill, but the energy quality of the food is less in a skinny, starving, sick animal. Because of the lower benefit, the wolf will have to kill and consume more often. Young animals are not as big, and the old and sick animals don't have the fat reserves of healthy animals and thus don't provide as many calories for the predator. In ungulates, the calorie content can vary around 400 calories per pound. Somebody figured out that a group of men starving to death in a lifeboat who eat a boatmate who just died of starvation don't get much if any nutritional value—few calories are still left in the corpse.

A few years ago in Yellowstone, a severe ice storm early in the winter covered the grass, making it less available to the foraging elk. Many elk starved or nearly starved to death, and the wolves scavenged these deficient elk with little or no transportation cost and little or no killing cost. The extra food, even though of relatively poor quality, was abundant, and the bears were hibernating.

The ice storm and the elk starving are illustrative of another significant point in behavioral ecology. For most animals the food supply is not constant from one end of the year until the next. Blooms and droughts within the seasons can render the food supply temporarily unavailable. In Yellowstone, the ice covered the elks' food. For some little period of time, they had no food. Elk are a ruminant, and ruminants don't endure starvation well, simply because they depend on a host of microorganisms in their rumens to digest their food. If their rumen is empty of forage for even a short time these microorganisms starve to death. When food becomes available again, the elk cannot digest it unless they can somehow get a new colony of microorganisms. Licking and sniffing noses with other elks is a good way to get a new supply of microorganisms.

The behavioral ecologist looks at this scene and notes that elk usually collect and store sugars (as fat and tissue) by eating plants. But because of the ice storm, hunting for grass became energetically too expensive for the elk. For many, digging through ice costs more in calories than they could get back from the little food they uncovered ($B/C < 1$). For the wolves, the transportation and the acquisition costs of hunting elk greatly declined, and they got many more calories than they needed just to survive ($B/C > 1$). They invested the surplus calories into pups.

These periods where food becomes unavailable—in this case for the elk—are called bottlenecks. It may be that for each species and each population, the ability to get through the bottlenecks defines how the many survival games are played. The long cold winter, the scarcity of food, and perhaps we should include disease—all of these put both individuals and populations at risk.

Often, when the population of wolves, coyotes, and jackals gets high, an infectious disease is more easily transferred from one animal to another—mange, for example—or in our Yellowstone story, the wolf pup population was afflicted with the usually lethal parvovirus. One way or another, wild populations increase and decrease quite frequently in the grand scale of life.

Plagues are good occurrences for those that survive them. When the population of coyotes, wolves, or jackals is high in any area, the competition between individuals of any species is high, resulting in troubles getting enough to feed on. Pup mortality will be high in any given year. Then the plague hits, and the population falls. At that point, feed will be more abundant per individual, competition between individuals will be low, and a higher percentage of pups will survive.

Cost of Feeding a Village Dog

Let's compare the foraging behavior of the village dogs with that of the wild canids. We'll start with an island population—that magical island known as the Mexico City dump. These dogs at the Mexico City dump (plate 5, *top*) are a good example of how the 850 million dogs outside of human reproductive control manage their feeding behavior.

Every day, the dogs sit in the Mexico City dump and wait for the garbage trucks to arrive. Their transportation cost to food in calories is about zero. Their acquisition cost is about zero. Since the food is dead on arrival, killing cost is about zero, and any repair cost minimal. In our wolf/moose story, for example, the moose doesn't exactly want to be killed and eaten and defends itself, whereas the discarded lamb chop at the dump doesn't care.

What about bottlenecks in the dump if food becomes short for one day? About zero. In other dog studies done in South Africa by Daniel

Stewart, the dogs are only eating once every other day or every third day. And, of course, that is true of wolves, which only get a bellyful once every three or more days.

What about seasonal fluctuation? About zero. People access a similar amount of food every day. They create a similar amount of waste from that food supply every day. The bottom line is that the scavenging dogs receive access to a food supply that varies little, every day of every month of every year. The garbage collectors have full-time jobs and work steadily every month of the year.

That should be the end of the story: a dump dog has few foraging costs compared with its wild relatives, which must put a huge effort into obtaining food. When one compares the cost with the benefit, dump dogs are way ahead. Getting calories is mostly easier for them.

Wolves are designed to feed in a particular way, having evolved to their wilderness niche. Their niche has piles of calories called moose, elk, or caribou, and wolves are designed for killing and eating big game. Usually we don't think of dogs as uniquely designed to feed in a particular way, like wolves or jackals. We don't usually think of them as a highly evolved species that lives in a niche where the transportation, acquisition, and repair costs to obtain food are exceedingly low. But we'd like to encourage you to think about *Canis familiaris*, the dog, in the same way you think about the evolution of wild *Canis* species.

The reason dogs make good pets is in large part because they have this innate behavior of finding somewhere to sit and wait for food to arrive, which is exactly what our pet dogs do. Their niche is scavenging food from humans. They are like ravens and foxes that scavenge food from wolves or humans (fig. 6, *bottom*). Where is that dog food supply? Look for humans, and there it is. Why are dogs nice to people? They are the source of food. Dogs find some food source that arrives daily and they sit there and wait. Being somebody's "pet" isn't all that different from being a dump dog or a street dog or a village dog.

One question we might ask, in order to find any foraging difference between dog and wolf, is whether wolves could scavenge the Mexico

City dump as well as dogs do. All of a sudden when we ask the question that way, we begin to realize how well designed the dog is for its niche. Remember the design costs. It takes 1,000 calories a day to maintain a dog, and it takes 2,500 calories for a wolf just to hang around acting doglike.

How about jackals and coyotes that are about the same size as our village dogs? Why aren't they in the dump, reducing their transportation, acquisition and repair costs? Interestingly enough, wolves, coyotes, dingoes, and jackals quite often scavenge in dumps, but only if the dumps are empty of people. It might have been that if it wasn't some species of the genus *Canis* that evolved into dogs the next best bet would have been the foxes. But that is just a guess. The tendency of the wild types is to avoid people, which, when you think about it, raises the transportation cost. But that is a different story, and it comes later in the discussion about the cost of avoiding hazards.

The avoidance of people, however, is going to give us a definition of a domestic animal: it is an animal that can eat in the presence of people. Dumps vary a lot, although scavenging in dumps has some basic similarities. Food arrives at a dump, on a somewhat continuous, even predictable basis. Families in the rural countryside often have their own refuse piles. In poor villages, people don't throw anything edible away. No matter how poor the quality of human food, it still might make a soup. Takings from single-family dumps might be so meagre that a single dog might have to scavenge the waste from several families. The figures aren't good on this yet, but extrapolating from observations in several African villages, data suggest that it takes the waste from a hundred people to feed seven dogs. This means that having a breeding population of dogs in town depends entirely on how many people live there and the quality of their waste. Many towns in the world don't have any sanitation units—which makes human feces available to the dogs.

The local outside-of-town dump dogs tend to live in small groups and seem to defend their dump from other dogs straying into the dump from town. The arrangement is a little unclear, because new dogs do join a group from time to time and individuals leave. The dogs in town tend to be spaced throughout a village, and individuals are usually in the same place every day, which implies that they are defending a resource.

Defending the resource will add to the cost. If the resource is like the Mexico City dump, it is impossible for a single dog or even a group of dogs to defend it. If the resource is large and concentrated like the dumpster out behind a hotel in Zanzibar, there might be three or four resident dogs that regard the resource as valuable and so will defend it successfully from other dogs with little energetic cost.

The Mexico City dump is a site into which a huge quantity of good-quality garbage arrives daily. Trucks dump piles of garbage in a progression that moves the front of the dump daily as the organizers fill up some valley or natural hole, and/or make a mountain. On our second trip, the Mexico City dump looked just the same as when we were there before. Then we realized it was in a different place. On the previous site a shopping mall had materialized. Dumps move continuously. They have a leading edge that moves daily—but also the whole dump can be situated somewhere else.

So much food typifies a moving front edge that it is impossible to defend it. Dogs, like wolves at a carcass, may argue but more often they share the site. When so much food is right there, what is the point of protecting it? In fact, defending too much food, for a single bellyful, is energetically and needlessly expensive.

In Chake-Chake, the main city on the island of Pemba, an abattoir is a favored place for dogs acquiring food. When the workers are slaughtering cattle, maybe one or two cattle at a time, a temporary abundance of waste is thrown outside. In a sense, it is thrown for the dogs because the dogs perform a service by cleaning up the place. Those dogs quite often fight over choice morsels. As a result, they are scarred and have open wounds. The repair cost incurred is great, and these dogs are often in poor condition. At this location, when food is present it is very rich in quality, but at the same time it is usually limited in quantity. Slaughtering cattle is sporadic and is not regular either in timing or quantity. And when business is slow, which it can be for days, the dogs disperse into the surroundings only to reappear when by-products of the activity are present.

Note an interesting lesson here. The timing of the food's appearance and the location of its appearance interact with the quality of the food to produce distinctive behaviors among the dogs. The quality of the food changes the behavior of those accessing it. Years ago, a research colleague of ours performed the following experiment on hummingbirds in Costa Rica. He hung a hummingbird feeder with a tube running in and out, which allowed him to easily change the contents of the syrup in the feeder—which was basically sugared water of varied concentration. When the concentration of sugar was low, the hummingbirds would share the feeder. But at high concentrations, the birds would fight over the feeder. Indeed, some hummingbirds would defend the feeder and not let the other birds feed. It was a controlled experiment in the sense that the location and the quantity were constant—but the quality changed.

We used this principle in managing our livestock-guarding dog kennels. Every once in a while, we would get too many dogs and would have to double up some dogs in the runs. That meant feeding two big reproductive males in the same pen. We had found with a series of experiments that pig pellets were a perfectly good dog food. After all, pigs and dogs are both simple-stomached mammals and require about the same amount of everything per pound. The dogs, however, didn't like the taste of the pig pellets. (They were used to delicious dry dog chow—which we often supplemented with food scraps from the college kitchen.) They wouldn't fight over pig pellets. Two big breeding males would eat out of the same dish.

If we threw in a mush of table scraps, we might have a dog fight. The same was true with a pen full of puppies. With high-quality dog food, the big pups would defend it. One pup would gorge feed and then lie in the food pan and not let anybody else feed. Often such a pup, which is getting too much high-quality food, would start to grow malformed legs or hips. Pups that were fed on pig pellets would eat many times during the day but never gorge and never defend the food from littermates. Think how important that is to the behavior of developing puppies.

Dogs are not the only scavengers on people that enjoy the low transportation and acquisition costs. Cockroaches, rats, pigeons, chickens, and vultures are a few of the others. Each is adapted to a different scavenging niche. Often the attraction to the food depends on its size: seeds, carcasses, and so on. Chickens and pigeons wander the streets picking up grain that somehow gets there on a regular basis. Even insects that are attracted to humans for whatever reason can become a food supply for some other species.

As we noted earlier, in the nineteenth century, house sparrows were a plague on the American landscape. Waste seed in horse poop was the attraction. The transportation and acquisition costs for the birds were almost zero. All they had to do was sit around and wait for a horse to deliver. It is the same model as dogs in the dump waiting for people to deliver the waste products of their economy. Dogs frequently visit latrines, waiting for people to deliver waste products, just as the sparrow does with the horse. People made all kinds of suggestions for getting rid of the sparrows, which were regarded as pests, fouling up farms and equipment with their excrement. At the time, *Smithsonian Magazine* even published a recipe for cooking them. The sparrow problem was eventually solved when the horse population declined. It may be that the dog population problem would be solved with a reduction of people.

An Old Story with a Twist: Fewer Vultures—More Dogs

One more example of the effect of human activity on the dog population comes from an unsuspected source. Vultures are scavengers. They eat dead animals, and in places such as India, many dead animals are old cows. Cows are another domestic animal, and the dead ones can be found near people. Other scavengers such as dogs could also eat dead cows, but the problem with eating old dead cows is finding them. As at the abattoirs, the cows appear randomly, but the difference is that the abattoir's location is constant. The presence of cows at the butcher shop is signaled through the countryside with an increase in activity at the abattoir and the mooing of the cows. With the dead cow in the field or roadside, the timing of the food source is erratic and similarly the location changes.

Biblically, dogs were known as the scavengers of the city, and vultures as the scavengers of the field. Dead cows are mainly in the field, the domain of the vulture. Vultures cruising ceaselessly on the wind or thermals can cover hundreds of square miles cheaply, looking for food. We have a vulture feeder in our upper field. It is just a platform high enough so our dog can't roll in the fish cleaned on the feeder or the road kills we collect. Fishing isn't an everyday event for some of us, so fish carcasses show up on the feeder only irregularly. Appearance of road kills depends on how much we drive, and whether it is convenient to stop. Thus, food in the feeder is sporadic (fig. 6, *center*).

The sky will be absent of vultures when the remains of the filleted fish go onto the feeder. But it isn't long—sometimes less than an hour—before at least one vulture shows up. If the fish is cleaned at night, a bunch of vultures shows up in the morning. (If it is winter and the vultures have gone south, the crows take over the job.)

If we were doing the behavioral ecology of vultures, we would look for our same formula $B/C > 1$. What is the transportation cost of the vulture finding something to eat versus the transportation cost of a dog finding the same food? In the field, vultures will find food more quickly and with a lot less energy than a dog. The vulture weighs three or four pounds, and our typical village dog weighs almost ten times as much. For the dog, searching the countryside for dead cows or dead anything adds a tremendous transportation cost. Realize that cows dying is an irregular event. The dead-cow food is not exactly the correct formula for dog food in that it doesn't come at regular intervals to the same place. A vulture floating hundreds of feet in the air can watch a huge country-side where they can often spot a dead cow somewhere.

Vultures have a similar problem to that we described for wolves. Once they find that pile of food called a dead cow, they have the prob-lem of protecting it. Often many vultures are attracted to a cow and end up gorging on it. As it is for wolves, they will eat all they can hold—but soon the other scavengers arrive. At our vulture feeder, on a good day, several of our bald, black friends will argue over the food, and some-times others are circling in the air above the commotion (fig. 6, *center*).

Then the crows arrive and also a coyote. It could be that the crows and coyotes were searching the countryside and happened upon today's goodies late, or it could be they are watching the vultures and are at-tracted by their noise and activity. Certainly, the vultures are attracted to the activity of other vultures. The coyote may have smelled the new treasure through the forest because smell will go around trees. More likely, they hear or see the commotion and are attracted to it. Many birds like terns will fly toward other circling terns, supposedly drawn by their feeding behaviors. In New Mexico, we often found sheep car-casses left by predators by watching the sky for circling vultures. Find-ing where and what the vultures are eating lowers the transportation cost for these other scavenger species.

Dogs also can smell, see, and hear the food-producing activities. Off the east coast of Africa, on the island of Pemba, we saw the fish-ing boats come in and the fish being cleaned and prepared for market. Dogs knew the routine. They would travel from wherever they were and concentrate at the site. Pickings were good. Now here comes the point of the vulture/cow/dog story. Presently in India, the vultures are being killed by a veterinary pharmaceutical used to protect cattle from inflammation and to give them some pain relief. The compound, a non-steroidal anti-inflammatory drug, isn't toxic to mammals but causes renal failure in birds that eat those treated mammals. The vultures are particularly prone to it because they scavenge those old working cows that have been kept alive an extra year with the medicine.

The population of vultures in India has plummeted, and some people fear they will go extinct. The reduction in the vulture population means piles of food are not being found and eaten quickly and efficiently. This gives the dogs a chance to make a living in the countryside. As a result, some estimates have postulated an increase in the dog population in millions, which is good news for dogs. They have found a niche that is becoming vacant and are rapidly filling it. The bad news is they are spreading rabies to the farming population, where it hasn't been prevalent in the past.

Evolution in Action

The crux of part 2 so far is to show how the parameters of evolution affect a population of mammals. The struggles of the predators and prey all show convincing illustrations of evolution in action. The village dog, which has taken advantage both of the low cost of designing an efficient size and shape and of access to food, has made itself at home in an environment that includes humans.

As we've seen, compared with the wolf, the dog has an immensely lower design cost. It only takes a few months to get that design operational. Wolves are larger and have a higher calorie-expensive design. It takes at least two years to get that design product up and running efficiently.

The dog doesn't have to run and chase in order to acquire sustenance. Compared with wolves, the dogs have a much lower transportation cost and a miniscule acquisition cost. We don't know about the repair cost. Is it as dangerous living in the dump or on a city street as it is trying to kill a moose or an elk? Nobody has studied that as far as we know, but most wolves on Isle Royale are dead by five years old. A look around the Mexico City dump suggests the mortality age of the adult dogs averages out about the same. What is needed are survival data and a comparison of the structures of each population.

The dog's job is to find a spot where food arrives regularly. Defending a spot depends heavily on the timing of the arrival, the quantity, the quality, and the distribution of the food. On any particular day, if the food hasn't arrived, it does not make much sense to defend the location of its future arrival. When food does arrive, the best way of defending it is to eat all you can. Fighting over an abundant supply adds needlessly to the acquisition cost. In places such as the Mexico City dump, comparatively little fighting arises among dogs, especially over food (plate 6).

The food-gathering process for a dog is usually relatively cheap, compared with that of their wild cousins. This niche, the human household, supports the village or street dog and even our household dogs. This population of canids survives better than any other canid species. That is the why and the how of the world supporting a billion dogs but only 400,000 wolves. The 850 million dogs not under human reproductive control are not strays or mongrels but are, in reality, their own unique, well-adapted species.

8

The Cost of Reproduction

When you look at reproductive behavior, you find the benefit/cost equation is different from that of design and foraging. The purpose of reproduction is to transmit genes into the next generation. It doesn't matter what this costs as long as it is done. In many species, it costs the parents their lives. And that is okay—the reward is huge if they are successful in producing reproductive offspring.

It is almost a rule for any species that reproduction is enormously expensive. That is what life, evolution, and natural selection are for. Successful animals repeatedly pay the price for their success, but it is well worth it in terms of survival of the species. Think of the old saying, "A chicken is an egg's way of making another egg." Then consider the design cost of producing a chicken just to make an egg.

In our previous discussion of the measurement of foraging, we emphasized the relationship of calories spent and calories gained. We saw that the behavior and the reward are immediate and measurable. Now, in contrast, we will see that the benefit of the reproductive process cannot quickly be measured. For example, a female codfish can produce 6 million fry per year, and all the predators in the ocean including codfish themselves prey on all those millions of little growing codfish. To continue the species, at least two individuals, male and female, are required to replace the parents. If these two succeed in reproducing, the population of codfish is stable and the parents contribute their genes to the next generation.

On average if a codfish has a 0.000003 percent success rate, producing 6 million eggs, then why not produce 12 million eggs and double the chances? That gets complicated. One problem is that it takes lots of energy to produce that many eggs, and it might not be feasible for the female to amass enough energy to make 12 million fry. Maybe it is better to produce 6 million good eggs rather than 12 million poor eggs with not enough yolk to sustain them. Not only the size of the mother but also her condition is crucial to success. (The same constraints pertain to the male.)

So why don't codfish evolve a bigger size so they can lay more eggs? A review of chapter 6, "Design Cost," will provide an answer to that question.

A second problem with more eggs is that those 6 million little codfish are competing with one another in the little codfish niche. Instead of each little fish competing with 6 million siblings, they now would have to compete with 12 million. The more little codfish the parents produce, the more each little fish will compete with one another for food in their baby codfish niche. Their chances for survival change for the worse with the larger competition.

Maybe the codfish that produce the most fry have the best chance, and yet 6 million is the best, the very best, that a codfish can accomplish in one year. Then maybe success depends not on how many fish per hatch but on how many hatches the fish can produce in a lifetime.

What is true for the codfish is true for the wolf—and for the village dog. If a wolf female has about five pups a year starting in her second year for the next three years (aged two, three, four, and five), then she has twenty pups in her lifetime. If two pups survive to reproductive age (one to replace her and one to replace the male), the population will be stable. The two surviving pups are just enough to replace their aging parents every five years.

To put it another way, in a normal stable population, 90 percent of the pups born will die before they get to reproductive age. In one study on Isle Royale, two-thirds of yearling wolves never reached five years old because of starvation-related deaths. In another study, on Yellowstone wolves, 50 percent of the pups were dead at eight months, and the surviving pups have another fourteen months and two winters before they are ready to breed for the first time.

Wolves, jackals, dingoes, and coyotes have relatively low mortality numbers compared with species such as codfish. However, that is not exactly the way to count. What is the purpose of reproduction? Our vulture illustration becomes important here. A species occupies a niche. The vultures of India aren't reproducing enough to keep their niche full. If it is not full, then something else moves in. When the vultures in India died off, the dogs and rats moved into the vacated niche. "Moved in" means that, in time, 5 million dogs inhabited the niche that had been exclusive vulture habitat. Suppose, then, that people stopped using the toxic cow medicine, and the vultures started to recover. Even though the dogs might not be quite as well designed for that niche, the recovering vulture population would have to displace 5 million dogs before the system came back to what it was.

In a perfect world, holding the reproductive cost at a minimum, the birth rate would be precisely equal to the death rate. There are several problems with that. The carrying capacity of the niche can change seasonally and over the years. In 2014, the populations of wolves on Isle Royale and in Yellowstone National Park were the lowest they had been in years. At the same time, the elk herd in Yellowstone was also the lowest it had been in years, but on Isle Royale the moose herd has been increasing. For wolves, a decrease in their prey base meant a smaller carrying capacity, but on Isle Royale, with the increase in moose, the carrying capacity should have increased. Obviously, many factors can determine the population number.

Ecologists consider two kinds of death: (1) density-independent death and (2) density-dependent death. The two can overlap.

Density-independent death is death by accident or some catastrophe. Perhaps it's a terrible storm with giant hailstones or a fire that kills everybody in spite of how many animals were there before the catastrophe. It doesn't matter how many animals are in the population or how full the niche is. A couple of years ago, our local chickadee population plummeted. Chickadees were suddenly so rare at our farm that we got out of our chairs to go look when a chickadee was reported at the feeder. What had happened was density-independent mortality: rain and cold and a general lack of bugs for two weeks in the spring when the eggs had just hatched. Few if any of the nestlings were fed enough to sustain them through that natural disaster. It wouldn't have mattered whether there were two nests or a thousand nests. Their parents couldn't find enough food or provide enough warmth to keep the nestlings alive during this feeding bottleneck.

We have already mentioned the ice storm in early in winter that covered the grass in Yellowstone, making it unavailable to the elk. Because of the ice, the elk were held at greater than usual risk of starvation and more vulnerable to predation by wolves. It didn't matter how many elk there were: none of them could get enough food.

Density-dependent death increases with a population's size. When the population is high or exceeds the carrying capacity of the niche, all begin to starve. What happens in these cases is that the food in the niche decreases, meaning it won't support as many animals and more of them will starve or die of starvation-related diseases.

On any given day, an animal is more likely to contract a disease if the population is high simply because the space between individuals is less, and contact between them is more frequent. Put another way, when the niche becomes crowded, starvation is more prevalent and starvation-related deaths are more frequent. For all the canids, and especially for those that are highly social, such as wolves and dogs, disease will run unchecked through the population. Diseases such as mange are constantly present in wolf and village dog populations and often reach epidemic proportions when the density is high and the animals are weaker because of increased competition at high population levels.

At the same time, the best method of getting more offspring into the next generation and keeping the niche full regardless of the density-dependent or density-independent death rate is to over-reproduce. Over-reproduction may seem contraindicated in a population where death is more frequent if the density is high. But over-reproduction is an anomalous concept. It really means keeping the death rate as high as possible. When the carrying capacity is limited, such as in the bowl full of marbles discussed earlier, the birth rate should be timed to equal the death rate of adults. If the birth and death rates were just equal, then the bowl would remain just full, and everybody would live happily ever after — as long as the bowl remained the same size. Therein lies the catch. Nothing ever stays the same, and the only way to ensure the bowl is kept full is to over-reproduce.

It seems that is the biological rule. If individuals in the population over-reproduce and the adult death rate is low, then very few offspring need to survive to replace the old and keep the niche full. If, however, periodic catastrophic die-offs occur, then excess pups have a better chance of surviving in a sparsely populated niche.

The formula is: a good survival rate of adults will result in a large mortality rate among pups, and a large mortality of adults then will result in a better survival rate of pups. Straightforward as that. And why? Simply because the adult is the adapted shape. Pups that are growing need the calories but they don't have a shape that will compete with adults. This is particularly true in the street and village dog world, and we will come back to this point several times.

How much over-reproduction should there be? As much as is energetically possible. How much is possible? Well, different species have different strategies, based on the characteristics of their niche, their sizes, and their niche-related problems. It is also important to think about the age structure of the population.

The characteristic strategies for reproduction by domestic dogs differ significantly, even dramatically, from those of their wild cousins (coyotes, jackals, dingoes, and wolves). These strategic differences fall into three categories: seasonality, territoriality, and parental care.

Seasonality

The first difference in reproductive strategy, seasonality, is simply that all the wild types are on an annual cycle (fig. 7, *top*). Wild canids mate annually, at a time that anticipates a once-a-year photosynthetic flush—the proliferation of edible resources. Not only does that flush appear at different times in the northern and southern hemispheres, but the intensity of the photosynthetic bloom changes with latitude as well. The twenty-four hours of sunlight above the arctic circle result in a tremendous sugar production for a short period of time, while the twelve hours of sunlight in the tropics from one end of the year to the next probably produces tons of sugar but with no seasonal flush. For animals in the north, spring is April and May. Daylight lasts longer, temperatures are warmer, the ice melts, and leaves emerge and start making sugar.

It is easy for us in the north to assume that the signal to start breeding is the increasing day length, which starts in December in the Northern Hemisphere and in June in the Southern Hemisphere. Our Massachusetts coyotes start their courtship behavior in December/January and their pups are born in late April.

However, the old expression that "correlation does not mean causation" should be heeded here. Many observations suggest that at least something else besides day length is triggering the annual reproductive cycle.

For example, in any given year, all the litters of coyotes in a region will arrive within a couple of weeks of one another—essentially synchronously. From year to year, however, there may be a month's difference between whelping dates, one year to the next, even though the light cycle is identical. On the other side of the world, all the Ethiopian wolves will give birth within a couple of weeks in any given year but will vary by a month from year to year. Further confounding the picture is that the Ethiopian wolf is right on the equator, and the day length does not vary much. The fact that all gray wolves in a region whelp almost simultaneously suggests that they all ovulated on a cue. If the cue was lengthening daylight, then all the wolves in a region would have pups on basically the same day every year. But that's not, as just mentioned, what actually happens from one year to the next.

Dingoes come into reproductive readiness as the amount of light each day is declining rather than as the days are getting longer. Why is that? Nobody knows. Some people speculate it is because they are closely related to northern gray wolves and stick to the ancestral yearly pattern, producing pups in April and May. But that is improbable. More likely, dingoes, like all the other species, adapt to having pups at a season when their pups have the best chance for survival. The trigger must be something else besides just day length. Maybe a better hypothesis would be that the dingo's annual cycle is timed to the onset of a wet or dry season that produces a photosynthetic bloom; they are responding to a different set of signals and have adapted to it.

For equatorial canids such as golden jackals or Ethiopian wolves, the signal to start reproductive activities could hardly be the changing light because, from one end of the year to the next, day length doesn't change by more than fifteen minutes. There isn't that dramatic change from twenty-four hours of darkness to twenty-four hours of daylight that occurs above the arctic circle. Also, no photosynthetic bloom is correlated with that fifteen-minute change. The cue for ovulation is more likely a change in rainfall patterns, which would create an opportunity for plants to produce more sugar.

The crucial point in responding to any cycle is to drop the pups when food is available — at a time when food is plentiful. At the time of birth, more mouths demand to be fed, and females are not going to be in the best of condition after pregnancy and during lactation. Thus the transportation and acquisition costs of food must be cheap.

Viewing as much of the research as we could find about reproductive activities of the canid family around the world, our charts told us (not too surprisingly) that all the wild types have an annual reproductive cycle. For our argument here that means that they have pups once a year.

Not only do females ovulate once a year, but males only produce sperm for just a short period every year. After the mating season, both males and females are sexually quiescent for the rest of the year. Their reproductive organs are inactive. The males' testes can shrivel to the size of peanuts. Male wolves, jackals, and coyotes do not produce sperm for most of the year. Females don't ovulate again until the next season. Everything, though, has its exceptions: it seems that if female dingoes don't get pregnant on the first breeding, they can recycle and ovulate again. That must mean the males stay sexually active beyond the initial breeding season (fig. 7).

Seasonal testicular quiescence in males is a good example of a design cost. If wild canid males can breed only once a year—because all females are in season only during the same brief season each year— what is the sense of having an energy-expensive set of testes producing sperm all year around? When females are not in heat, the benefit of producing costly sperm is zero. It is an excellent example of how precise the evolutionary process can be. All individuals are out there competing for limited food resources. If some individual is using fewer calories by not producing sperm, then that individual should have the selective advantage over those that produce an expensive tissue pointlessly. The domestic dog is very different. The street dog and the village scavenger do not have annual cycles. The domestic scavenger is not faced with seasonal blooms and busts.

Humans store grain, which they can eat in the same amount every day over the entire year. They can store the grain in shelters or store

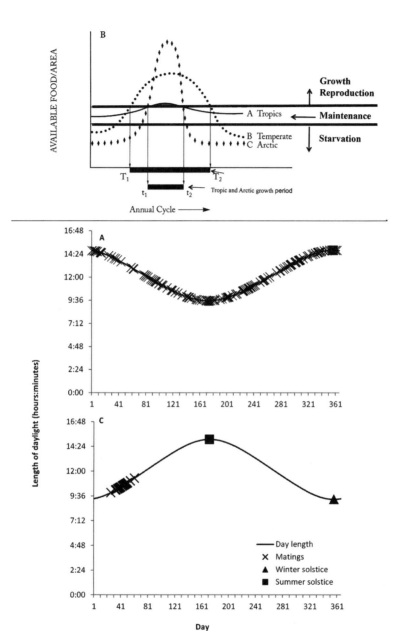

Figure 7. *Top*, Photosynthetic blooms differ from the equator to the arctic. Many species are adapted to having their offspring at the beginning of the bloom. (Adapted from Geist, *Deer of the World* [1998]). *Center* and *bottom*, Contrast the yearly reproductive periods in dogs and wolves. Wolves have one breeding season a year, while dogs can have pups in any season. (From Lord et al., "Variation in Reproductive Traits of Members of the Genus *Canis*" [2013]).

it in cattle growth. They produce waste products fairly regularly and evenly all year around. They tend to live continuously beside their food resources, such as coral reefs, grasslands, or grain-storage areas. They tend to their food supply on a yearly basis, growing both their own food and feed for their animals, stacking it up for the off-season, and protecting it from theft.

Humans tend to eat in one place day after day at the edge of their food supply. Often, their supply is stacked right next to where they consume it. They often accumulate their waste outside where it is accessible to wild animals — and dogs. If they live in a great big village — a town or city — they might hire somebody to take their wastes to a communal dump. They take it to the dump regularly, where the dogs are waiting for it.

Dogs have no need to wait for the photosynthetic bloom in order to breed and produce pups. They can breed any time. And they do. All the records we have ever seen say they breed any month of the year. Because female dogs are reproductively spaced throughout the year, their pups are competing with only one-twelfth of the annual pup crop. All the wild types of canids have their pups once a year and generally synchronously within their species. If in any given year that time turns out unseasonable, unusually cold, wet, or whatever, they could lose the year's crop of pups. Because all the newborn pups are on the ground at the same time, a disease going through at that time could destroy them all. In 2005, in Yellowstone National Park, a pup virus devastated almost the whole wolf pup crop. In those cases, the cost of all that courtship behavior, setting up territories, being pregnant, digging a den, and so on, amounts to zero benefit.

Territoriality

Dogs tend not to have feeding territories, and they court repetitively all year long, which leads to rich and robust social lives. More than almost any other mammal species, the dog's reproductive social behavior is continuous throughout the year.

Unlike their wild cousins, dogs are promiscuous. What looks like a pack of roving dogs is often a dozen males courting a single ovulating female. And they all have a chance to produce pups with her (plate 7).

A litter of dog pups could possibly have as many fathers as there are pups. That gives the female greater genetic diversity in her offspring, which should increase her chances of parenting a survivor into the next generation. For the male dog, instead of being limited to five pups per year if it is monogamous, it is possible to sire dozens of pups per year. In contrast, the wild-type male needs to have a territory and to pair-bond with a female, thus increasing the courting cost and also limiting the number of individuals in the population that can participate.

The wild species commonly do not generate sexually active females until they are two years old. The smaller coyotes and jackals can breed in their first season, but it's important to realize that in their first courting season they are only ten months old. In order to breed, a female has to reach a reproductive body size and has to be in reproductive condition. That is almost impossible for gray wolves, which are not full size yet in their tenth month, especially in the middle of their first winter. The smaller species — coyotes and jackals — can reach full size and have found a territory at ten months if all the stars have lined up for them. It is not uncommon, however, to find some of the jackals not having their first litter until they are three years old.

Remember that, in the discussion on foraging costs, we said that those big northern wolves are not even good hunters until they are two to three years old. This is partly explained by the fact that they haven't reached their final adult size. If the adult size is what the species has been selected for in order to be an efficient hunter of big game, anything less than the adaptive size is not as efficient — not as good at hunting or reproduction.

In all the wild species of the genus (meaning wolves, coyotes, golden jackals, black-backed jackals, side-striped jackals, Ethiopian wolves, and dingoes—for starters) not all of their females have pups in a given year. They can be too young or too old or perhaps shy of their reproductive weight or otherwise not in condition to bear pups. In some cases, a female will have pups one year and not the next partly because she hasn't recovered from last year's pupping and attendant activities. With the wild types, if they have not reached that minimum body condition in anticipation of the reproductive season, they won't ovulate, and as a consequence, they will miss a whole year.

The female dog also must reach some minimum body condition in order to ovulate. The good news is female dogs can ovulate any time they meet the minimum requirement. Many of them reach sexual condition at seven months.

Indeed, dogs can come into reproductive season more than once a year, depending on how you define a year. One female dog we know in South Africa had two litters totaling eighteen pups in the 2012 calendar year. Maybe she didn't have any pups in 2011 and maybe she won't have any in 2013 — and, then again, maybe she did and maybe she will.

To say that a dog has two heat cycles a year is not the same as saying they can have two litters a year. Normally, having two litters on the ground within twelve months isn't quite possible. Pups hit the ground (call it day zero) followed by fifty-six to seventy days of nursing, then 180 days of recycling and getting up to the body condition where ovulation can take place, and then another sixty-three days for the next litter to hit the ground. Ten or more months have elapsed. A dog can come into heat every six months provided she doesn't have a litter. It is possible for a dog to produce a second litter within eight months, but village dogs don't usually do it. It is uncommon for a dog to start to recycle while she is nursing, although a few can start right after giving birth. That may have something to with genetics, but more likely it has something to do with body condition. Nursing is expensive and many females lose fitness during that period. Think about it: she has to eat and convert the food into dog. Then she has to convert dog into milk. It is the old stomach-size problem again: Can she eat enough of good enough quality to make those conversions? Thirty-pound dogs give birth to half-pound pups that are the same weight as pups of much bigger dogs. At the same time, the thirty-pound dog does not have the stomach size or the body mass to produce as much milk—all other things being equal, which they never are.

Compared to the wild types, having a litter in ten months is almost a 20 percent increase in pups per year. Data on litter size for village or street dogs are rare, but it seems that many in our observations had seven or eight or more pups, which is another 35 percent increase in pups over the wild types.

Is that good? Well, it is different. Each species evolves a reproductive strategy based on a number of environmental restrictions. Figure 8 (*top*) illustrates those different strategies.

Parental Care

The third important difference in reproductive behavior between wild canids and domestic dogs is the length of time the parents care for the young. The exciting fact about wolves, jackals, coyotes, and dingoes is that they seem to exhibit the most highly evolved, complex parental behavior in the carnivore world. A group of South American "foxes," according to our South American colleagues, also show complex parental behavior. But to date we haven't found anything written on those South American distant cousins of our *Canis* wild types.

Most coyote and jackal females cannot raise five pups by themselves. In order for the male to convey his genes to the next generation, he needs to help raise those pups. In other words, if coyote and jackal males didn't help, the litter wouldn't survive, and neither the male's nor the female's genes would be carried into the next generation.

For the little foxes, each mother can catch enough mice all by herself to feed the growing pups up to some stage when they can hunt by themselves. Since a vixen usually has four pups in a year, it is to her

WOLVES – COYOTES – JACKALS

	♂	♀	helpers
NURSING		X	X
REGURGITATION (after weaning)	X	X	X
PROVISIONING	X	X	X

DOGS

	♂	♀	helpers
NURSING		X	
REGURGITATION (after weaning)			
PROVISIONING			

Figure 8. Reproductive strategies. The wild *Canis* species (*top*) have the most complex parental care behaviors in the carnivore world. Among dogs (*bottom*), males have nothing to do with puppy rearing, and females pretty much abandon pups after weaning.

benefit to put some quality time into taking care of them. For the male fox, the best thing to do is to try to get more than four pups per year into the next generation — thus he cruises looking for other females to mate with.

If both parents can't raise the litter, then it is a selective advantage for older siblings to "help" with raising pups. Wolves are the most commonly mentioned with this pack-feeding-the-pups strategy. Having an older pup that hasn't dispersed "helping" raise pups is also common in coyotes and jackals. Similarly, packs of wolves are commonly family groups in which last year's pups have not dispersed yet.

"Help" is in quotes because they may or may not actually be helping. Why haven't they dispersed? One answer is that they aren't big enough to be either sexually active or successfully hunting on their own. They are still hanging around their parents for access to food. They still need the help of their parents in finding game. But once the parents make a kill and they all gorge themselves, last year's pups can regurgitate some of the meal to this year's pups.

However, if it is a poor foraging year, the older pups may be competitive with their siblings for food regurgitated by the parents. They are still growing and could well be competitive for available food with their younger sibs. If it is a bad year, then the competition from the older pups could lead to the death of the younger pups. Provisioned or regurgitated food is eaten first by last year's pups. Believe it or not, the reproductive job of the parents is to survive genetically. It's important to raise at least two pups into the next generation; it doesn't matter to them from which litter they came.

The wild cousins of the dog are a phenomenon in how much care they give to their pups (fig. 8, *top*). The female nurses the pups after they are born and she will bring them food as they grow. What is totally unusual about the genus is that the female regurgitates food to the pups. What is even more unusual is the males' participation in providing calories for the next generation. The males will feed the female after she has denned up for the birth and while she is nursing pups. Then, after the nursing period, the male will feed the pups by regurgitating food to them and provisioning them with food—for months.

For the female, producing pups and nursing them for eight weeks is expensive and exhausting. She might need three or more times her normal calorie intake to nurse a litter of five. A female coyote or jackal normally needs 1,000–1,500 calories a day, and when they are pregnant, their requirement likely increases to 2,000 calories and while nursing as high as 3,000 calories. A seventy-pound female wolf normally needs 2,500–3,500 calories a day, which increases to 5,000 calories while pregnant and 7,500 calories while nursing. While she is pregnant and then nursing she is not in the ideal adaptive shape to be chasing and killing moose, elk, or other big animals, and thus her abilities to gather food has decreased.

The transportation and acquisition costs of collecting food for pups are expensive for her. The pups are up to a year or more away from being ready to feed themselves, regardless of which wild species we are considering

Dogs have evolved a different parental strategy (fig. 8, *bottom*). Human waste tends to show up at the same place daily and so the dogs, as we have noted, have very low transportation and acquisition costs.

The pregnant female village dog can stay by her food source all through pregnancy and lactation. She can locate her den in the middle of the food source (fig. 9, *top*). Frequently, she goes to some quiet place outside the village (but not too far outside). There are many quiet places in the Mexico City dump. All over the dump are fat nursing pups.

At the den, no male dog is anywhere to be seen around pups. One finds males by looking for females in season. The male's job is to impregnate as many females as possible at any season of the year.

Among dogs, there is no necessity for a feeding territory or pair bonding. Males have no contact with the pregnant female in the sense of regurgitating food or provisioning her. For males, then, it is not a selective advantage to care for pups that could well not be his own.

Indeed, because females are promiscuous, there is no guarantee

Figure 9. *Top*, These South African dog puppies live in an elaborate, well-constructed den. They are fat and well cared for. *Center*, Some scrawny juveniles try to compete with a well-fed, well-adapted adult. *Bottom*, A young juvenile, illustrating that the way for a pup to survive is to get adopted by a human. (Photos by Daniel Stewart.)

that she is pregnant with any particular male's pups, even if she had mated with him. Most litters are likely to be sired by more than one father, again underlining the selective disadvantage for any male in taking care of other males' offspring.

Pups can be nursed for up to ten weeks. The actual length of time depends on how long the female can produce milk, which is commonly about six weeks. Weaning takes place for several reasons. Pups would continue to nurse as long as the female would stand for them. She, however, tends to get sore teats, especially as the volume and quality of the milk decreases for the growing pups. Bigger pups need a larger volume of good-quality milk, which for the most part the female can't supply after that six weeks.

Here is where the greatest difference between the dogs and the wild types appears. Neither the male nor the female dogs supplement the pups' diet while they are nursing—after weaning the pups are on their own (fig. 8, *bottom*).

Everyone who has raised dogs will claim to have seen at least one female dog regurgitate "once" to their pre-weaned pups. We have surveyed hundreds of people at lectures and always someone has observed a regurgitation event—and we have, too. But nobody ever claimed it was more than an occasional observation or that it was significant in raising pups. Some reports in the literature note apparent pair bonding and/or males regurgitating to pups. In the category of captive/restricted dogs, researchers include several reports of apparent provisioning of pups. But as far as we know, no one has claimed that a litter was raised to maturity by attending parents.

In our experience, after weaning, fat little dog pups are on their own. They may follow the mother back to the village or to her feeding station in the dump, but at eight to ten weeks they become competitive with the adult dog population for food and everything else a dog needs. Is it possible that there are exceptions? Yes, of course, it is possible.

Pups' growing bodies may need just as many calories as the six-times-bigger adult. Their little stomachs and a poor quality of food mean they have to eat multiple times to obtain enough nutrition to grow. They aren't like the adults, which can gorge feed once every two days.

If the adult size is the ideal size—the adaptive size and shape—for a village dog, then the five-pound pups cannot compete effectively (fig. 9, *center*) with adult dogs. They usually don't do very well. Most of them starve. In the Mexico City dump or in a small city in Ethiopia, one finds the age structure consists mainly of nursing pups and adults.

A small adult population, perhaps because of some kind of die-off, gives these growing pups a better chance for survival. Humans are still producing the same amount of waste during and after the die-off; the remaining dogs, including juveniles, can do reasonably well.

Another strategy exists for juveniles to survive after having been weaned: being adopted by a human family (fig. 9, *bottom*). More about that in part 3, but to conclude this chapter about reproductive strategies, we'll repeat the idea that female dogs are like cuckoos, which lay their eggs in a strange nest to be cared for by someone else. Pups adopted by humans are cared for and have a better chance of surviving than those that aren't.

9

Avoiding Hazards and Their Costs

All creatures are mortal. The best strategy for survival is not to die before you have left your genes to the next generation. Biologists think of life expectancy as a generation time—with each generation passing its genes to the next. As we have shown, most wild individuals' genes don't make it to the next generation because of a lack of food and calories to get them to reproductive age. Most adults are narrowly on the edge of an adequate food supply. Most deaths result from weakness related to not having enough food. That is all normal.

Most species watch out for dangers continuously. Prey species have to be constantly on the alert for predators. In most species, the young are vulnerable to predation. Wolf, coyote, and jackal pups are vulnerable to big cats, big raptors, bears, snakes, and a host of other animals. Pups have to be hidden and guarded. Building dens, moving pups to new locations, and remaining vigilant are all energy expensive.

What isn't always apparent is how much energy goes into being aware of hazards. One year, we watched, way out in the field, a wild turkey, and her dozen chicks. They were all searching through the grasses for whatever little turkeys eat. The female, however, was spending most of her time watching for danger. We would see her head above the grass, looking, looking, and looking, in all directions, all the time. She was constantly nervous. She would steer the chicks away from open spots and at the slightest possibility of a threat she sounded her special quiet alarm call and the chicks would dive for cover and stay motionless. By fall she had four full-sized chicks left. Even though they were full size, it was easy to tell which one was the mother, because she was mostly skin and bones.

The four chicks had a design cost. It took many calories to build them into full-sized adult turkeys. Yet in neither of the previous chapters (chaps. 7 and 8) on costs did we mention hazard avoidance. In the chapter on reproduction, we had to add in the calorie cost for parenting offspring. Now we have the calorie cost of just being continuously aware of possible dangers and protecting the young from threats to their lives.

Turkeys are what we would usually think of as a prey species, although the thousands of insects they eat would probably not classify them that way. Many so-called prey species, such as elk, moose, caribou, or bison adults, are only marginally prey unless they are weak or sick or young. Wolves are most interested in the young of those species, which for the wolves have cheaper acquisition and repair costs. Young ungulates have few defenses other than protection by their mother and other members of their social group. Each species has evolved a weapon, such as horns or defensive behaviors, to protect the young. Bison will form a circle, facing outward, with their calves in the middle. All those defensive measures are noteworthy, but most of us seldom add these to the cost in calories of surviving.

Canid wild types make or find a den in some out-of-the way place.

The females are alert. Sneaking up on them and catching them asleep is a rare event.

One of the hardest parts of guarding young or even of adults protecting themselves is to stay alert and protective while searching for food. Most animals have to take a chance when feeding. Part of the transportation cost for any of the *Canis* species is staying alert to potential dangers while en route. It is called being nervous—and nervousness costs more calories.

Staying alert to potential problems differs significantly in a variety of ways for wolves, coyotes, and dogs. Again, it is the dogs that are the most different in their behavior. Dogs are not as afraid or as nervous around people as the wild canids are. If you approach dogs in a dump or in an African village, they won't start to move until you get really close. Then they move only a short way and slowly. In fact, they will move in a wide circle such that they end up at the food pile again. It is very neat to watch. They move away unhurriedly (not many calories) from the approaching person while still guiding their movement toward the food.

In a study of dogs in an Ethiopian village, Alessia Ortolani, a stranger in the village, would walk past sleeping dogs on the street. In many cases, the dog didn't even move. Others would simply walk away.

Years ago, we created a project in Minnesota to record wolf vocalizations. Our method was to be all set at a feeding site before dusk. When the wolf family came in, everything had to be ready—before they arrived on site. One tiny movement, or the clicking on of a tape recorder, and they would be gone. Not only gone but not coming back today. Think about it: they ran away from food. Being safe is more important than food.

Thus, we can measure hazard avoidance the same way we measure foraging behavior. These are additive, as with the transportation cost, to the cost of survival. The animal runs away and that can be measured in calories. How far do they run away? The farther they run, the more calories it costs. The faster they run, the more calories. In the old days, on a cattle drive, a stampede—running because frightened—could run tons of pounds off a herd. Recently, it was found in Wyoming that cattle growth, in areas where there had been predation by wolves, was slower. Calves would be twenty-five pounds less at market time than their unstressed relatives. The assumption is that these animals are putting energy into staying alert and avoiding being preyed on. It is no different with elk or moose or turkeys or coyotes, wolves, and jackals.

One might think that shyness is a developmental trait. Dogs born with people all around would not be afraid of people when they grew up. Dogs born away from people are spooky around them when they grow up. At the same time, socializing wolves to humans by hand raising them is a long process, which means maybe 2,000 hours of contact when they are baby pups. Any of the wild species' pups start their social development with their noses—before their eyes and ears are open or functional. Wolves and coyotes don't socialize with humans or other species the same way dogs do.

Even when you socialize the wild types with people from twelve days old, however, you don't get a dog.

Exceptions do occur, but they are limited. Red foxes have moved into many cities, and they are reasonably tame around people, but not to the point of lying around and sleeping on the streets during the day. We raised coyotes in the same pens from the same young age as Border collies in order to record any differences in behavioral development. At weekly intervals when they were young—during the first year—we would take scales and rulers into the pens and weigh and measure both species. It was hard to keep the Border collies from jumping all over the equipment and the researchers while doing the work. The coyotes could be caught and measured, but it certainly was more difficult because their tendency was to avoid the researchers. Our coyote × beagle crosses were neither coyotes nor dogs. The hybrids were shy like coyotes, even though they were raised like dogs. You couldn't approach them and put a hand on them. Our tamed coyotes were all the same, in that no matter how much time you spent with them they did not behave like dogs.

We have pictures of a (wild) female black-backed jackal that visited a ranch house in South Africa and watched the dogs and mated with the ranch's male Border collie. She produced four pups in the surrounding area but they were shyer than the mother. In Quebec one summer, wild wolves came to our hand to feed. All our fellow wolf experts yelped when they saw the pictures and called us foolish simply because it was a dangerous thing to do. It is one of those situations where we all have a story, but all agree the dog is behaviorally a different animal no matter how you raise the wolves.

Whatever the stories, the village dog spends less energy than a coyote or fox while feeding in and around the village. The dog's flight distances from someone approaching are much less than those of the wild canids; their flight path away from somebody is much, much shorter and slower moving. All of which saves energy.

The point, however, is that there must be calories for the nervous energy all animals put into being watchful. Additionally, there must be spare calories stored in case the animal has to run away from a danger. If a bear arrives while the wolves are feeding and takes over their kill, the wolves then have to hunt another prey, expending extra transportation and acquisition costs without have gotten much benefit from the first. They must have enough stored energy to run away from the bear, or try to defend the carcass, or go on another hunt. If they lose two carcasses in a row, what then? It isn't as if they could just go to the gas station when they are low on energy and fill up.

The dogs waiting for food at the dump are relatively calm; they are adapted — evolved to eating in the presence of people. That is why dogs make good pets and the wild types don't.

PART III

That Special Relationship between People and Dogs

10

The Symbiotic Relationship

The earth does not shake when we recognize that dogs and people share a special relationship. Certainly, with our pet dogs, many of which are distinctive breeds we paid good money for, the pride of ownership is important.

More significantly, of course, there's that happy dog, always glad to welcome you home. People enjoy such a "special relationship" with dogs that, throughout the developed world, the word "pet" has become almost politically incorrect. "Pets" have become "companions," and we do not think of owning them any more than we would think of owning a child, even though we pay all the bills.

There are at least two problems with the viewpoint that dogs are companion animals, pets, or "man's best friend." The first is that the overwhelming majority of people on earth do not think of dogs as companions, to be owned and paid for. It is not that they think about it and reject the idea. Rather, ownership does not represent the kind of relationship most of the world's people have with dogs. The idea of caring for a dog daily by remembering to buy dog food in advance, or keeping its vaccinations current by taking it to a veterinary office, or walking it on a leash to the dog park so it can get proper exercise is not how most people in the world view those dogs in the yard. Indeed, in many East African villages we have visited, people would claim ownership of the chickens running freely around the neighborhood, but not the dogs. "That is my chicken!" while the dog barely has a name. The chicken has value and the dog does not.

Is that your dog? Yes. Does he have a name? No. Is that your tree? Yes. How do you know that is your tree? It is in my yard. Does your tree have a name?

For a biologist, the second problem with the reputation of the dog as companion animal is that characterizing it thus does not tell you what the selective advantage is of this relationship between dogs and people—or, why it is "special." Is this special relationship biological? Has this special relationship evolved over time, as some researchers claim? People and dogs, they conclude, are adapted to live together in some special relationship. Having a dog is good for you and leads to a better life or having a dog leads to a longer life. Maybe having a dog leads to not only more children, but better quality children, who would grow up and have more children. Does a dog help convey more of your genes into the following generations? Now that is something that would give people who have dogs a selective advantage over people who do not have dogs.

For biologists, two species living together is described as a "symbiotic" relationship. That simply means that they live in proximity to each other and perhaps interact in some way. Yet labeling it symbiotic doesn't reveal what the selective advantages of living together are—or even whether there are such advantages.

Living together might simply mean that their niches overlap. Prairie chickens nest on grasslands and bison graze on the grass. Sometimes a bison accidentally steps on a prairie chicken nest and breaks all the eggs. For a population biologist studying prairie chickens, this would be a factor to be researched. The reproductive success of the prairie chicken might be inversely proportional to the population size of the bison. The denser the bison population, the more accidents will occur, and so the worse for the reproductive success of the birds. Yet it doesn't change the quality of life for the bison at all. That type of symbiotic relationship is called amensalism by ecologists.

Biologists have names for all the different kinds of symbiotic relationships. For many people, the word "symbiotic" becomes synonymous with a mutualism in which living together benefits both species. "Benefit for both species" means, to the biologist, that people who have dogs will have more and better children than those that don't and dogs that are helping people in some way will have more and better puppies than those dogs that live independently of people.

The classic illustration of mutualism is hummingbirds and flowers. The flower has evolved a specific nectar-producing structure for which the hummingbird has evolved a distinctly designed bill to get that nectar, which is the basis of its diet. At the same time, the flower is fertilized by pollen stuck to the hummingbirds' uniquely shaped bill.

In this case, we call the symbiotic relationship mutualism because both species benefit. However, the hummingbird-flower connection is more than that. It is an obligatory mutualism. No hummingbirds — no fertilization of flowers; no flowers — the hummingbirds starve to death. The population of flowers increases, and the population of hummingbirds goes up.

Often, as we talk about the relationship between dogs and people, we imagine the relationship is a mutualism — something good for both species. That mutual relationship is illustrated well with working or hunting dogs. One theory says dogs were domesticated because they helped people hunt. The biological implication is that those people who had primitive hunting dogs gained better reproductive success and had more or better children than those people who did not have hunting dogs. Evolutionary biologists might test the hypothesis that those people who domesticated dogs to help them hunt had a selective advantage over their neighbors. They might test whether the domesticated dogs had a better chance of reproductive survival than their wild ancestors did. They would test whether those wolf ancestors of dogs that helped people hunt experienced better reproductive success than those wolves that did not help people.

This sort of scenario on dog evolution—starting with no data—allows the seekers of dog origins to be very imaginative. Imagine sled dogs working in teams pulling sleds full of seals, which could be used for both human and dog food. The claim is for a mutual benefit for sled dogs and humans because the dogs that pull sleds have a better chance of reproductive success than their nonpulling cousins. Those people who worked with sled dogs also increased their chance of reproductive success because of the reduced transportation cost in acquiring food.

It is a great scenario, which illustrates the survival benefit for both species and advances the hypothesis of the mutualistic "man's best friend." However, are there data to support the conclusion about the dogs pulling sleds of seal meat? Such scenarios remind us of Rudyard Kipling's *Just So Stories* (highly fantasized stories of the origin of some animals).

Put the bits of the story together, and one is led to more and more speculative conclusions. For example, the Siberian husky today is promulgated as an "ancient breed" of sled dog, tracing its ancestry all the way back to the dawn of people-dog relationships. The Siberian must have been the dog that reproductively outcompeted all the other ancient Siberian dogs. This dog and the people who lived in Siberia must have evolved symbiotic mutual benefits living side by side with hunting-gathering Chukchi Eskimos. These Mesolithic preagricultural-period people and their dogs have existed together for thousands of years, and the AKC Siberian husky is proclaimed as a genetic and physical representation of that ancient relationship.

Actually, there is no evidence for any of these assertions. If native North American or Asian peoples used sled dogs before the thirteenth century, they left no evidence for us. No evidence does not mean, however, it did not happen. Interestingly, the first evidence of sled-dog teams comes from the Thule Eskimos at about the time they encountered the Vikings, though there is no evidence that the Eskimos learned how to drive dogs from the Vikings. It is probably just a coincidence that the evidence of sled-dog teams and encountering the Vikings were concurrent. This might make another great story.

Consider an alternate scenario: the symbiosis between humans and dogs is verifiable because wherever you find dogs you also find humans. If we wish to compare this relationship to our hummingbird-flower situation, we need first to contemplate a number of factors regarding the latter. If we get rid of the flowers, hummingbirds are in serious jeopardy.

Now ponder those same scenarios with people and dogs. If dogs disappeared from the face of the earth today, would it make any difference at all to human reproduction? Some people would be sad for a bit, and some people would need to go into a different profession. But nobody has made the argument that the loss of *Canis familiaris* would have much effect on human reproduction. Indeed, the vast amount of evidence is that, among the world's population of dogs, most are pests and spreaders of disease. If dogs disappeared, then 70,000 people would not die of rabies from dog bites next year. More people would get a good and uninterrupted night's sleep. And so on.

If humans disappeared from the face of the earth today, dogs as we recognize them would go extinct. Why couldn't they go back to the wilderness and hunt rabbits or moose? Because wolves, coyotes, jackals, and foxes already occupy those wilderness niches, and those species already adapted to those niches would outcompete the dog. Those wild critters have the right shape and instinct, and the dog would have to evolve into something else. One niche — one species.

Dogs and humans do live in a symbiotic relationship, one that ecologists call commensalism or eating at the same table. For dogs, it is obligatory. The niche the dogs have evolved to is one in which they can make their own living only from human waste. No humans—no human waste.

The question might arise as to whether this obligatory commensalism could have started with preagricultural hunter-gatherers. Didn't those Mesolithic people—hunter-gatherers some 30,000 years ago—shed human waste? Why couldn't there have been evolving dogs scavenging the carcasses of game with those hunter-gatherer peoples?

It is a common belief (though not ours) that dogs evolved from wolves—and the implication is usually that those were the big gray wolves. When one looks at the kinds of skulls that anthropologists claim might be the first dogs, they are always the hundred-pound gray, northern wolf size. So what is the possibility that Mesolithic hunters domesticated the dog from those big wolves? We were involved for years with Wolf Park in Battle Ground, Indiana. They had about fifteen wolves at any one time, which they fed mostly on road-killed deer. The feeding formula was five pounds of deer per wolf per day. That is about 3,500 calories (715 calories per pound of deer) per day.

Let's apply that data to the common theme about the origins of dogs—that early man took wolf pups from the den and domesticated them as hunting companions—and then do the math. If wild wolves need about 2,500–3,500 calories a day and don't really start to be good hunters for two years, then the average Mesolithic hunter had to put 2.19 million calories into every growing wolf cub before it went on its first companion hunting trip with the human and symbiotically caught their first moose together. That means the early domesticator has to kill a deer once every eight days to feed a single growing pup. That is a lot of work, and it yields only one pup. One pup does not a reproductive population make. A human raising a wolf cub does not make a dog, either.

Well, what if these wolf pups were eating the waste from Mesolithic hunting camps? We've calculated that it takes the waste from a hundred people to feed seven dogs, or fourteen people per one dog in temperate African villages. Consider that today's thirty-pound village dogs need less than a thousand calories per day. The wolf needs three times that number of calories and therefore would require the waste from three times as many people, or between thirty-five and forty-two people per wolf.

How many dogs does it take to make a reproductive population? The quick, slick, and incorrect answer is two — one male and one female. That is wrong simply because the reason for having a male and female is to have pups (let's say six pups a year) that live long enough so that one male and one female offspring can reproduce in the next generation. Now you have eight dogs, which need the waste from over a hundred people.

Let us guess that the minimum reproductive population has twenty animals in the various age classes (which to some of us still seems low). If there were wolves running around the Mesolithic village, it would take the waste of 800 Mesolithic people to support them. It is hard to imagine Mesolithic hunter-gatherers running around hunting and gathering for groups of 800 people. This scenario is beginning to sound reasonably unreasonable.

Maybe we have the wrong view of Mesolithic hunter-gatherers. Maybe the emphasis is on gathering and not so much on hunting. Good evidence shows that many populations of these preagricultural people lived in caves long enough to draw pictures on the walls. Their cousins sat on a shell fishery for generations. Thus, it is possible to imagine a situation where the wild type could evolve into the obligatory commensal dog. The exercise for the reader is to imagine the numbers involved.

The special symbiotic relationship between dogs and people comes with noteworthy design changes in the dog and none in people. This is a good indication that it is the dog that evolved into this obligatory commensal relationship, not the people. The design changes are not only physical but behavioral as well.

The design of the dog is significant when one realizes that the niche is/was human discards. For the wild canid, it becomes a requirement to be able to eat in the presence of humans. In order to eat human waste and compete with other animals that are after the same waste, the animal has to get close to humans as they shed those wastes. Most important, the scavenger must not run away when humans appear. Therein is the dog that we all know and love, the species that hangs around people and waits for something edible to fall within reach.

Wolves can and do feed in the modern dump, as noted earlier, but not as well as dogs. In many of our studies, we found that watching wolves at the dump is tricky. The minute the observers are discovered—the wolves are gone. In a sort of "beam me up Scotty" reaction, the wolves instantly evaporate when they perceive a human nearby. Even worse, for our observations, you might not see them again for days—if ever.

More serious for the wolf that runs away is that it leaves the food resource exposed and available to a competitor. Look at the dump material as if it were a wolf kill that has been abandoned. It's now available to other scavengers. Watch crows or vultures at a kill—they will retreat in the presence of an approaching person but they don't go far, and they return quickly once the person is gone. The blue jays at our bird feeder are like that. They are instantly gone when they notice any motion in the house. However, they have not gone very far.

As you approach dogs in the dump, to recap what we said earlier, you note that they do not move until the very last minute and then they move away slowly, reluctantly almost, and in a circling path that brings them back to the food. The adaptation revealed is the very short flight distance, the slowness of the withdrawal from the resource, and the circular pattern of withdrawal. None of these behaviors is energetically expensive, and they keep the animal focused on the resource and not the escape (hazard avoidance). Resources are limited, and if the wolves abandon them quickly (which takes energy), for hours at a time, whenever a potential hazard such as a human comes to the dump bringing more waste, it is going to be tough to make a living there. If the wolves feed only at night, when humans are not coming and going, then wolf scavenging is energetically more productive. The best-adapted scavengers are going spend the most time at the food source. That is the dog.

Ultimately, on the subject of evolving this symbiotic relationship with humans, all the other rules of intraspecific competition are extant. When you watch dogs in a city or village, you see they are competitive with one another. They are so well adapted to this niche that other species give them very little competition. Between dogs, however, competition for the available resources can be serious. Necessary resources such as water are in short supply in the Mexico City dump. In the top photo in figure 10, an almost imperceptible threat of the black-and-white dog moves its neighbor away.

Protecting dog resources from other dogs is significant, despite the infrequent occurrence of actual fighting over food. Intimidation is very much like other evolutionary traits, which are based on a low calorie budget. The body posture is often the maximum of the investment. In the bottom photo of figure 10, you see one dog close to the woman who is cleaning the innards of a cow prior to making sausage; the dog's proximity to the woman and the food, and the turn of its head (the "look") toward the other dog, are signs of a threat to the other dog. The other dog is intimidated from approaching the food, as you can tell by the turning of its head away from the first dog and the distance at which it is sitting. No fight required.

The ability to move close to the source of that waste as it is being produced is a selective advantage. Close means closer than other dogs. Two components achieve this closeness: (1) getting the human to allow access to the leftovers, and (2) intimidating other dogs out of that space.

The resource is limited: it takes a hundred people to feed seven dogs. And which seven dogs are going to get fed? The ones best adapted to the scavenging niche.

Figure 10. *Top,* Water is a valuable resource. Here the whites of the eye of one dog show "the look," causing the younger dog to turn away. (Photo by David Muriello.) *Bottom,* One dog intimidating another dog with just the "look" is enough to protect the space.

The Bad Side of That Special Relationship

In addition to dog owners and professors of evolutionary biology, governments are also interested in the symbiotic relationships of dogs and people. The United Nations World Health Organization (WHO) is very interested in dogs. Dogs are a pest species for most local and national governments for many reasons, the worst being they are the vector for several human diseases (fig. 11, *top*).

Rightly or wrongly, WHO (at least the field representatives representing WHO, with whom we talked in Ethiopia and South Africa) feel that dogs are not only the vector (the distributor) of rabies but also the reservoir, much as humans are the vector of AIDS and also the reservoir in that one can only get AIDs from another human. Why WHO considers dogs to be the reservoir we do not know. It may be there are different species of rabies and one is a specialist on dogs. We could make a different argument, but they are in charge and their policy is to monitor dogs closely and to try to manage and limit the negative effect dogs have on people. In WHO's defense, given the adaptation of dogs to feed in the presence of humans, dogs infected with rabies are common in many places where they are in intimate contact with people — maybe more so than any other species. That in itself has made them a special problem since the beginning of dogs.

In Massachusetts, we vaccinate our milk cows against rabies. Mostly we do that to protect the cows from the disease but also to protect the milkers and others who come into contact with their fluids. Any mammal can get rabies and transmit it to people. The dog is just particularly good at it.

Because one of WHO's jobs is to manage dog populations, it has studied the various relationships of dogs and people and devised the following four categories of dog-human relationships:

1. Restricted dogs (fully dependent and fully restricted)
    These dogs are fully restricted in their movements. They are,

Cost to Humans

1) Dog bites
    world wide – millions and millions/year

2) Rabies
    world wide – 70,000 people get rabies from dog bites

3) Other diseases
    echincoccosis, hydatidodis, toxocarosis –who knows

4) Livestock killing and harassment
    estimates vary by country but $ millions

5) Negative effects on wildlife
    spreading of disease, hosts of harassment problems

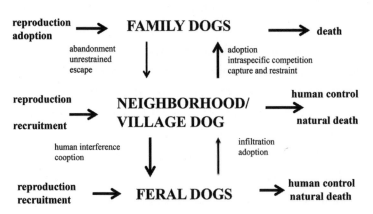

Figure 11. *Top,* Cost to humans: problems with the man-dog symbiosis. *Bottom,* Luigi Boitani and his colleagues categorized the world's populations of dogs. The emphasis of his study is that dogs can change status one or more times during their life.

for the most part, the pet dogs in so-called first world countries. The restricted category might also include some percentage of working and hunting dogs. Because their movements are totally in human control, they are totally dependent on humans for food and other necessities. Humans control even their reproduction. Fully restricted might also mean they are neutered and incapable of reproducing.

Most governments, humane societies, and welfare groups tend to believe that all dogs should be in this category. Indeed, many humane organizations go to countries where dogs are not fully restricted, capture them, neuter them, and relocate them into environments where humans control all their movements.

2. Family dogs (fully dependent and semirestricted)

These are dogs that are dependent on people feeding them but are only semirestricted in their movements. It is a difficult category to define. "Fully dependent" means the dog is fed by humans and has no other access to food. It might mean that humans control the reproductive behavior of their dog or it might not mean that. At the same time, the dog might be able to wander in the neighborhood and therefore has a better chance of spreading disease or exhibiting other pestlike behaviors, such as chasing cars and bicycles, biting people, fighting with other dogs, and barking all night, to just mention a few. However, what "fully dependent" actually means and what "semirestricted" means are difficult to define.

In our rural community, no garbage lies around, and our sanitary facilities are indoors. Our dog is fully dependent on commercial dog food. Even though we have leash laws in the town, our short-legged Jack Russell has a small dog-operated door (push with nose to go out, push with nose to come in) and can go in and out whenever he wants. Most of the time, the dog stays in our yard. In his old age, though, he started to make rounds in the neighborhood, and some of our neighbors fed him from their dog food bags or table scraps. Often, he helps himself from water dishes on their decks. Our dog is neutered and vaccinated against rabies. He is a family dog, meaning totally dependent on humans for food, totally in our reproductive control, though he is semi-restricted.

Many family dogs around the world are not neutered and/or vaccinated. They may be totally dependent in the sense that they are given food, but they run their own reproductive lives within the larger community. In the United States, half the litters born are said to be "accidental." That means that the family thinks it is controlling reproduction but actually only has a 50 percent success rate. The fact that the dog roams means that anyone being bitten by a dog needs to get a rabies shot because there is no guarantee that the dog has been vaccinated unless the owner can produce a certificate.

3. Neighborhood dogs (semidependent and semi- or unrestricted)

These dogs are semidependent on humans and semi- or unre-stricted in their actions. Since our dog can go in and out anytime and gets food from our neighbors as well as at home, maybe he is not just a family dog but has become a neighborhood dog later in his life. In our observations, it seems that most of the dogs in the world are neighborhood dogs. Most places do not have com-mercial dog food. Even where it is available most "dog owners" do not see feeding a dog as their duty and they cannot afford to buy commercial dog food. Most "dog owners" could not afford to neuter "their" dog even if they had access to a veterinarian, nor do they have any of the mechanisms for preventing unwanted lit-ters. The idea of having a fenced yard and a method for keeping a female in season isolated is almost ludicrous. Those of us who try to run breeding programs are always amazed at the cleverness of dogs when it comes to breeding or being bred.

4. Feral dogs (independent and unrestricted)

These dogs live independently of people and unrestricted by

them. The term "feral," meaning wild, untamed, undomesticated, and hence savage according to Webster's dictionary, seems a bit exaggerated for these dogs. No dogs (almost) live in the wild. Shy dogs move into dumps outside villages, and their feeding activities are performed at night. After dark, they might venture into villages, but for the most part they avoid people. Yet they are still domestic, living from human waste.

Also, fully independent and fully unrestricted dogs have adopted a close relationship with people. The dogs in the Mexico City dump feed totally by themselves and are totally unrestricted in their reproductive activities, but they do hang around with dump workers. Some of them have names and will often come when called. Some go home with the workers at night, and some do not. They fall into that category of totally independent and totally unrestricted dogs, but many people seem to love them, talk about them as pets, and don't think of them as feral.

Maybe there should be a fifth category, which would include the dump dogs. But that doesn't help us with the exceptions in each of the other categories. It is difficult to know whether to classify our own dog as a restricted dog most of the time and a family dog some of the time. Sometimes he is a popular neighborhood dog. Maybe someone at WHO would recognize that our dog, at the same time as he fits the domestic categories, could become totally independent of individual people and totally unrestricted, and thus be classified as a feral dog. He has been sterilized, and in a sense is reproductively restricted.

Why would one think that dogs would be born into one of those categories and remain there? If one wanted to work out a management plan for vaccinating dogs, one would need to know what the age structure and the reproductive rate of the population is. As we reported earlier, humans have no control over the reproductive activities of something like the 850 million dogs that live in our world. The age structure of this population was figured at 50 percent nursing pups with a 4 percent survival rate into the adult class, which is the other "50 percent." Trying to keep such a population vaccinated given a high mortality rate of juveniles, is difficult, expensive, and not very rewarding (plate 9).

Luigi Boitani and his co-researchers worked out a variation of WHO's system, dividing the dogs into three categories and looking at reproduction and recruitment in each. The diagram in figure 11 (*bottom*) is an adaptation of the original graph in very minor ways. Boitani, for example, did not think there was much reproduction in the feral dog population. In three years of study in rural Italy, there didn't appear to be much reproduction because pups rarely replaced losses within the feral population. That doesn't mean that some 4 percent of the pups didn't find their way into the neighborhood or family dog population and survive there undetected. We will discuss this later, but in many areas, pups will search for a provider in the villages. It is one of those cluttered problems with not many data.

In his important study, Boitani categorized the three discreet study populations as family dogs, village and neighborhood dogs, and feral dogs (fig. 11, *bottom*). He did record some migration between them. In many of the locations we studied, we found that migration could be on a daily basis and, for any single dog, might change over and over again with time (plate 8).

Many who write about these categories assume that the family dog is at least a pet dog while the village/neighborhood dogs are strays, street dogs, or as our Namibian friends called them, pavement specials. The experts assumed they had been pets but were now abandoned. That assumption is partly based on the mostly friendly behavior of the dogs. They sit on the streets or sometimes in people's yards. We don't think they are abandoned pets, however. We think they are dogs that have evolved and developed good people skills. Historically, this behavior shows not only how dogs are but also how dogs are supposed to be.

From our observations, the dogs labeled as feral—wild with little or no attachment to people—are for the most part shy dogs. There are several kinds of shy dogs. Some were barely socialized with people during their critical period of socialization (during their first fourteen weeks or so of life). They could, for instance, be sheepdogs that were socialized with sheep but not people. Thus they could be not shy of sheep and terribly shy of people. When they are lost from their flocks, they end up feeding in dumps at night when people are not there and are labeled as feral dogs.

There is another kind of shy dog, which seems to have a genetic disposition for shyness. When we were raising sled dogs, we called these dispositionally shy dogs "kennel spooks." We would have two female littermates growing up in the same pen, and one would turn out to be a spook while the other had a normal dog disposition. Genetic shyness was common in our sled dogs, but it was not a fault in a good racing dog. They would race past a crowd of people, never looking from one side of the trail to the other but straight ahead. That genetic shyness was also prevalent in some of the Maremma sheepdogs. We were often warned away from some kennels when we were buying sheepdogs because of the high frequency of spooks, though spooks are actually useful in some situations. In an area of high human activity with joggers or bicyclists or hikers traveling past dogs guarding sheep in a pasture, the passersby would not be harassed or even barked at by the spooks. That in itself cut down the liabilities of having a dog in such a protective situation. The spooky dog would stay with the sheep and not intimidate or bite passersby. It seems as if the very few feral dogs we have seen are just these sort of shy dogs.

If the dog is in the house, it must be a pet and a family dog. That is an assumption. We found dogs in houses in Ethiopia and Turkey, which were not pets nor were they family dogs or even neighborhood dogs. They actually go into the house with the livestock that are kept there, usually in a separate room or on a lower dirt floor, while the people live above. In pastoral societies, these village dogs become bonded with the livestock during their critical period of socialization. They are also scavenging from people and even from the livestock, cleaning up afterbirths and carcasses. Several authors have tried to distinguish between pet dogs and others, assuming those "others" are not pets or working dogs. Those "other" dogs are seen as uncared for, products of irresponsible ownership, or otherwise aberrant. Those assumptions are based on the presumption that humans domesticated the dogs and therefore have a responsibility to care for them—all of them, correcting the supposed injustices of their street life.

In our research, we questioned people with dogs in several areas of the world. We, along with the students working with us, asked several sets of questions that we had committed to memory. When the opportunity arose, we asked carefully. When we got to a village, people would find out we were interested in dogs, which produced diverse reactions. A few of the village people who also liked dogs would make an effort to talk with us. However, other people who liked dogs and who distrusted us would seem to look for our hidden motives. They might proceed ahead of us, hiding particular dogs so we could not capture them and take them away. This was true in the Mexico City dump where one of the dog-loving workers, called Grenais, who supplied us with ever so much information, said quite out of the blue one day, "You won't betray us will you?"

One time in Turkey where people were not supposed to like dogs because of their religion, we were looking to obtain pups for our livestock-guarding dog project. We offered to buy Anatolian shepherd dogs or their pups. A religious man told us that if we needed a dog then God would want us to have it and we could just take as many as we wanted. As we moved through the town looking for pups, occasionally we saw people ahead of us frantically capturing a favorite pup and hiding it from us. We did manage to come away with three great pups that went on to protect American sheep from coyotes and other predators. They became founding stock for a breed new to America.

The Mexico City dump is a wonderful place to ask questions about the special relationship between dogs and people. Remember that the dogs there are fully independent, have a constant supply of food, and are totally unrestricted in all their movements, including their reproductive lives. They can court and breed and can leave and return anytime they want. By everybody's definition, they should be classified as feral dogs.

The problem with the feral dog definition is that the dump is also full of people. Two hundred people may work there, recycling materials. Small groups of them have actually moved into the dump with their families in order to protect the products of their labors. During the day, they will bag glass or plastic bottles, cans, pieces of metal, and tie up cardboard. These collections are stacked, and trucks come periodically to pick them up. They are valuable and thus need to be protected from theft, day and night.

It is quite a community of people. One young woman runs a childcare business, giving donkey cart rides around the dump to young children while their parents forage.

Years earlier, we had made a comparable study of the Tijuana dump, which was similar in most ways but at the time we thought those dump workers were poor people forced to the edge of civilization because of their poverty. However, when we read the studies about dumps worldwide, we found that many if not most large dumps support a similarly active labor force that recycles materials. In the Mexico City dump, the workers were making three to four times the minimum wage made by those who buy their recycled materials. Most of this labor is highly organized to the point where the Buenos Aires dump workers have gone on strike; also, the Mexican workers are asked to participate in political rallies.

From a dog point of view, city dumps and rural dumps differ. Dumps can be loaded with resident dogs, as in Mexico, or made up of more day migrants, such as in Tijuana. Some dumps have active programs to keep dogs out. The main city dump in Madrid has no dogs because of a huge fence around it and active removal of any dog that gets in.

The Mexico City dump is relocated from time to time as sites fill up.

The first Mexico City dump we ever studied has a beautiful shopping mall on top now, and the new dump is a couple of miles away. What is interesting—really interesting—about the Mexico City dump is that when they move the dump they move the dogs with it.

So here is the point. About 700 dogs live in the Mexico City dump. They are totally independent and totally unrestricted. Reproductively, people do not control them at all. Yet people care about the dogs and make sure that when the dump is moved, the dogs move with it. The dogs seem to care about particular dump workers and follow them around the dump.

We taught a weeklong research course in the Mexico City dump. One exercise was to interview the dump workers about the dogs. Each of the students plus some skillful translators memorized the questionnaire. One of the students who translated for us was Eliza Ruiz Izaguirre, who went on to earn a PhD at Utrecht University with a thesis titled "A Village Dog Is Not a Stray."

Remembering that the dogs are feral dogs, independent, and unrestrained, we asked, "Do you have dogs?" Fourteen of the people said yes, and eight said no. With such a small sample size, do we believe that two-thirds of the dump people would claim dogs? Yes.

Do we believe these numbers are representative of the 200 people there? When you ask these same questions in African or South American villages, you are questioning someone with a dog close by and you have no idea whether it is a family dog, a village dog, or a feral dog. The interviewer asks, "Is that your dog?" Yes. Or even though the dog is sitting in the doorway of the house, the answer can be an emphatic no. It reminded us of the old Peter Sellers joke: Does your dog bite? No. And then after the dog bites Sellers, he says "You said he didn't bite!" To which the reply was, "But that is not my dog."

One reason for the diversity of answers, of course, is that people lie. People are often suspicious, at first, of your intentions. One woman said incredulously, "You came all the way to Portugal to talk about dogs?" Alternatively, you could be the dogcatcher, taking all unowned dogs away, and so the person says yes to save the dog. Or, the person thinks you are going to tax dog owners and says no, even though they think of the dog as a family pet.

Next question: How many dogs do you have here in the dump? Of the sixteen people or families asked, they claimed on average four dogs each. Keep remembering that these are feral dogs, and everybody agrees that they are not fed or restricted by the people that claim them.

Favorite question: Does your dog have a name? Eighty percent of the people said yes. A small number of people said no. And then there was a third group that said some of their dogs had names.

Does your dog have a name? When we started asking these questions we assumed that if a dog had a name, it was evidence that the dog was at least somebody's pet. However, that did not turn out to be true either. One time a shepherd, all alone in a great mountain pasture with his sheep and dogs was asked, "What are your dogs' names?" He gave an amused smile and said "Let's see, you can call that one Cani [dog]." Many times when we interview people like this shepherd, we find that he did not even know the dogs. He was a hired shepherd for a big commercial outfit and he was assigned a new flock of sheep along with its dogs. With livestock dogs, one often finds that the dog and sheep are a unit, and the shepherd is a migrant worker from another country who has no interest in the dogs. Livestock dogs often switch flocks, and a shepherd whose sheep they are will smile and say he had never seen that dog before. Some dogs can be attached to a single sheep and go where that sheep goes, or in other cases the dog is attached to the shepherd and will go with whatever flock he goes with. The point is that some of these dogs did not have a human they were attached to. At the same time, they are not strays, or even products of irresponsible owners, or unfed, or uncared for. One would not ask, for instance, if the sheep in the flock had names.

In villages, people would sometimes give you a name almost embarrassedly as they realized you would expect the dog to have a name. Quite often a name was made up on the spot, such as, "We just call him Dog." Even so, the dog was a pet, but with no name.

Many a street dog has no "owner." Is that your dog? No. Does he have a name? Yes, we call him Spot. Occasionally, directions are even given, based on a dog: "Go down the street and when you come to Spot, turn left." Thus family dogs might have names, and at least thirty-two dogs in our feral dump collection had names. Several times in South Africa, we found dogs that scavenged from several houses. At each location, the family would claim the dog and tell us its name. One dog had five names, each associated with a different family, and was still classified as independent and unrestricted.

Another question we asked in the dump: Where do your dogs stay at night? Twenty-four percent stayed with the owner in the dump, while 41 percent went home with a worker. Only 12 percent of the dogs stayed in the dump when their worker went home.

Question: Why do you have (dump) dogs? Six people claimed protection, three said they simply like dogs, and three just felt sorry for them. One man would collect and sell pups born in the dump. Quite often in villages, people would respond that dogs were good protection. When asked "protection from what?" in East Africa many people said, "From other dogs." Some people had a relationship with their "feral" dogs, giving tidbits to arouse the dog's territorial instincts to keep other dogs away. At a hotel in Zanzibar, the owners hired the local veterinarian to neuter and vaccinate three dogs that were living on their beach. Why didn't you just kill them? Because they would just be replaced by three more. These three "volunteers" were named and fed hotel scraps, which is to say that they had first crack at the daily scraps that were thrown out. The three dogs were successful in protecting this feeding hot spot from other dogs. Note that if the three dogs protected this hotel from unknown people, it would be a problem for the hotel managers. These three, however, were reasonably spooky and tended to avoid attention from people. If you ignored them, they would ignore you. Still, they acted like pets of the hotel, sleeping in the shade of a beach chair.

For the most part, the dogs we witnessed around the world were peaceful with people and with other dogs. Fighting between dogs in the villages and the dumps was very low. In the dumps, there was some fighting over water (fig. 10, *top*). In most dumps, including the Mexico City dump, water is often a limiting variable, and dogs are given water by the workers or they have to leave the dump and search for water. When dogs were given water in the dump, we noted some aggressive posturing between dogs actively drinking. Seventeen percent of the workers said that the dogs fought over food, while 58 percent said that sexual access was the main cause of fighting, and 25 percent said they fight to defend territories.

We saw dogs aggressively defending shade, which also appeared to be limited and treated as a resource. One might see ten dogs lying under a truck or cart for the shade. Access to shade and water increased the incidence of threatening behavior. One individual displacing another from shade or water accomplished this by threatening but not fighting. Access to females in heat also increased the frequency of threatening behavior.

However, the dogs seldom fought during the day over anything. Even growling was at a low frequency. A simple posturing, showing the whites of the eyes, seemed to be enough to warn off other dogs (fig. 10, *bottom*).

The definition of territory was also interestingly different. Individual dogs were located in their usual spot day after day, suggesting they occupied a territory. When a garbage truck arrived, the dogs from all over the dump went to the dumping site and searched in groups, communally, for food. Again, we saw surprisingly little aggression. It seemed that the territory is a resting spot, maybe with water and shade. Similarly, the breeding of females by males may or may not have anything to do with territory. Usually, a female in heat traveled around the dump and the resting males bred her as she traveled past. Or a group of males followed her from place to place, and several males would breed her, serially. They would stand in line and wait their turn. Still, little actual fighting occurred, which is not to say there was not aggressive posturing. After breeding a female, a male would wander away looking for some place to nap.

Nighttime is a different story. Of the thirty people we questioned, 20 percent stayed at the dump with their families, often in tents. The purpose was obviously to guard their collection. If they stayed in the dump and an intruder showed up, the dog(s) would bark—thus providing protection. Seven percent of the respondents stayed overnight sometimes. If the collection trucks showed up late in the day, then the laborer could go home that evening. Seventy percent of the workers lived nearby and went home in the evening.

This is remarkable primarily because the claim was that the dogs are different at night. One hundred percent of the people we interviewed said that not only did barking increase but fighting among the dogs also increased. A high percentage of people (80 percent) said the population of dogs increased at night. We could not confirm this, because we could not go into the dump at night because at night the dogs (and maybe people) became more aggressive to strangers. Over 90 percent of the people we asked formally, and also our guides and bodyguards, said it would be foolish to go into the dump at night.

Did we believe them? Yes. When we were doing the study in Ethiopia, the field-workers provided no data for nighttime activity, believing it would be foolish to go out at night "because the dogs will kill you," reported the study leader Alessia Ortolani.

In KwaZulu-Natal, we investigated several deaths of humans killed by dogs at night. There was "always" something else going on. The person had been drinking and maybe had lain down or the person was handicapped and had an odd limp or a mental affliction.

Going back to the question of having dogs as protection, these nighttime aggressive dogs added an important twist. In our Ethiopian villages, nobody went out at night because of the dogs. Therefore, people felt that nobody would come to their house and hurt them—certainly no stranger. Thus, they felt protected by the dogs. "The dogs" meant not just their dogs.

Certainly, the concept of dogs being more hostile to strangers requires some discussion. What constitutes a stranger? Once, at an Italian shepherd fair, we saw a bunch of "sheepdogs" on display tethered together in what could only be described as a gaggle. It was obvious to our students and ourselves that we should not go there, or try to touch one. At that point, an older man walked into the group of dogs, not particularly affectionately using his staff to nudge them here and there. Are those your dogs? No. Can we come in there with you? *No*—they will bite you. Why don't they don't bite you? Because I smell like a shepherd. And he did, too.

In other places, we got similar answers, with people telling us we did not smell like the locals. A stranger it is claimed is someone who smells different. A person coming home drunk might smell different. Certainly the long history of teaching war dogs to discriminate between races of people could be based on smell. The Spanish, during the conquest of the Americas, purposely trained dogs to kill and eat the natives.

Most of the people who live in the dump believe that the population of dogs increases at night, although a few think it remains the same. This was difficult for us to sort out. Given that many dogs go home with the workers, how could the population go up? Maybe it's just that the dogs are more active after dark. Maybe because they are barking and fighting more, it seems as if the population has gone up. Concerning the many dogs that went home with the workers, perhaps they did so to avoid the increased aggressiveness at night.

There is, however, another possibility: it may be that a different feral population moves into the dump at night. Such dogs might be reacting to the reduced human activity at night, which might be enough to motivate shy dogs to come in and feed. If two populations of dogs are, in fact, using the dump, that might explain why barking and fighting increase at night.

In the dumps outside of towns in Italy, it did seem that shy feral dogs became more evident at night. In several places in South Africa, the question arose as to whether dogs active at abattoirs were coming from neighboring towns or were part of a group hiding during the day in a nearby forest.

The major point here is that trying to classify dogs in broad categories such as family dogs or neighborhood dogs or feral dogs is difficult because many dogs change categories during their lifetimes. Many change from the start of the day to the end of it, but wake up tomorrow back in yesterday's first category. Quite possibly, two different kinds of feral dog populations operate in the Mexico City dump, one during the day as sweet lovable animals with names and a home life, and the other a group of nighttime nasty and spooky marauders. If the nighttime marauders were actually just unsocialized or genetically shy dogs, then they would interact normally with the permanent resident dogs. "Normally" might mean aggressively interacting over food and reproductive activities. As a human, you might not have any trouble with them no matter how you smelled simply because they are avoiding people.

The special relationship between people and dogs is something the dogs are initiating. Dogs are involved in an obligatory commensal symbiotic relationship with people. It is for the dog's benefit and reproductive survival to enter into a relationship where they can feed in the presence of humans. They can accomplish those tasks in many ways.

The rest of part 3 elucidates the way that the symbiotic relationships are established.

11

Dogs Adopt People (and Other Animals)

Quite often in the United States, efforts are undertaken to urge people to adopt dogs from shelters. However, in our experience around the world, it is quite often the other way around, in that the special relationship between people and dogs is initiated by dogs. Indeed, for survival and reproductive success of young juvenile dogs, the trick for a dog is to get itself adopted.

When we asked Mexico City dump workers where they got their dogs, fifty-one of the fifty-four people we asked got them from the dump, while two bought their dogs somewhere else, and one was given a dog from somewhere else. Among this small population of people going to work every day, 70 percent go to an off-dump home in the evening. Some of them are followed from the dump by many dogs in the late afternoon, and followed by dogs back to the dump in the morning. The people work for money all day, and the dogs feed and reproduce at the dump all day.

That was the same pattern we saw in the Tijuana Dump years before, but at that time we didn't realize the significance. The behavior of the dogs might make you think that the dogs in the dump had belonged to local people and were dogs that had been abandoned or had strayed from an occupied home. One argument against this conclusion is that Mexico City dump dogs are uniform in size and shape. They almost look like a breed. They all look alike.

Mexico City has a population of people that really like their purebred dogs, and they promenade a rich assortment along the popular paths daily. The dogs at the dump do not look like a random assortment of household pets or breeds that have strayed. They look like the thirty-pound yellow-tan, black-and-white, short-haired village dogs that are ubiquitous around the world. This suggests some form of selection is occurring, meaning natural selection for size, shape, and behavior.

The other problem with the stray dog hypothesis is that the pups are born in the dump. Indeed, litters of pups live all over the dump. Those little sites where you see individual dogs every day often provide shade and shelter for fat nursing pups. In the dump, there are three kinds of female dogs: those that are getting pregnant, those that are pregnant, and those that are nursing pups. Again we are using the Mexico City dump dogs as illustrative of the other 850 million village and street dogs that spend a vast amount of their time getting pups on the ground.

The impression one gets when studying the Mexico City dump is that the hundreds of adult dogs you see there look good. The hundreds of nursing pups tend to look fat and content. But juveniles are practically nonexistent. When you see any juvenile pup, it looks awful. A careful search of the dump shows dead juveniles that have crawled underneath something, looking very much like they starved to death.

Therein is one of the big secrets of dogs—all dogs: their mothers abandon them at weaning and start to cycle back into heat to produce another litter. At the moment of abandonment, which means for whatever reason she stops nursing them, the pups have two choices: either go out there and compete with the adults for the food resources in the village or dump or adopt a person. In the Mexico City dump, it's hard for them to get adopted because people are not feeding dogs. Thus the juvenile is unlikely to follow a worker home since it doesn't seem to offer any particular advantage because the worker is not feeding the dogs.

Juveniles do not compete well with the adults, which have the adaptive shape, size, and behavior to scavenge on human waste. The young juvenile will not reach that adaptive shape and size for another five or six months and is not competitive for limited resources in the interim. A growing juvenile might need as many calories as the adult but does not have a stomach big enough to get enough food value out of the low-quality garbage. If the niche is full, meaning it has reached its carrying capacity, then the chances of successfully competing with the adults is poor, and mortality from starvation-related causes is very high. If, in contrast, the adult population is below the carrying capacity, because they were just killed off by a disease or calamity, then the pup has a better chance of surviving. We are back to the Darwinian axiom that all species over-reproduce and only the fittest will survive.

The question is, How does the nursing pup make the transition from a fat neonate to a reproductive adult? The answer is that as soon as the mother stops nursing, really their best shot at surviving is to adopt a human: soliciting care from some receptive person. After all, the dog niche consists of human waste, and the best of that is going to be near humans. It helps if the human thinks the pup is cute.

In various countries around the world, dogs are doing just that. The dogs themselves are adopting people. Watching a Maasai village, or a Zulu village, or the Mexico City dump and interviewing people about "their" dogs is illuminating about how people just seem to acquire dogs — but from where they just don't know.

We were talking to a worker woman who was being followed by a large number of adult dogs on her way home from the dump. We asked the usual questions, beginning with how many dogs she had. She replied, with a big smile, fourteen. However, one of them was new that day. She pointed it out, and it was a perfectly good adult male village dog, which had just joined her gaggle of dogs. It did not have a name yet. It was common to see dump workers followed by "their dogs." They would be amused by the line of dogs behind them and they knew the line varied from day to day (plate 10, *inset*). Sometimes the dogs followed them around the dump during the day. Some dogs had more than one name, depending who they followed home. It was the same in the mountains with the sheepdogs that followed different flocks on different days or the village dogs that scavenged at several houses.

We talked with a Zulu lady and her new baby (plate 10), asking: "Is that your dog?" She replied yes. "Does it have a name?" She giggled and said yes, but hesitantly. "Then do you feed the dog?" we asked. She giggled and said, "Yes. He eats in the backyard"—meaning the trash pile and the latrine. If she really liked the dog, she might throw something into the garbage that was not quite garbage yet, but it is essentially the waste from the family's table that the dog eats. In many African, Asian, and South American towns, the family does not have its own latrine. In many places, outhouses and latrines are not dog proof. The dog gets a dietary benefit from these frequented places.

We then asked the Zulu woman, "Where did you get the dog?" "I don't know." It was not something she had done but rather something the dog had done: the pup had simply shown up—it adopted her. It had solicited care from her, and that is how it survived to become an adult. It sits there at her house or, occasionally, at a neighbor's house. Is it a feeding territory? You bet. Does it protect her? Yes, from other dogs.

A neighbor of hers was feeding some pups, and we bought one from her. In KwaZulu-Natal, the pups are considered beautiful and often referred to as Africanis—the African dog. When we ask her, "Where did you get all these pups?" she sighs and says, "Ah, they just keep coming. I wish there were not so many." In the top photo in plate 12, a woman in Lesotho and her little son feed a favorite pup. The other pup, at right, is starving to death.

There are a thousand examples of this phenomenon. Pups in every village solicit care from somebody. Pups will come to where some other dog is being fed, trying to get a little something.

As discussed in chapter 9, the first big difference between dogs and their wild relatives is that dog sires and dams don't take care of the juveniles—they are on their own. The second big difference is that dogs have a long critical period of socialization compared with the wild canids. For dogs, this window of social learning extends from about four to fourteen weeks of age. This allows them time to continue their social bonding after they are weaned. In contrast, species-recognition patterns in wolves, coyotes, and jackals can be in place at three weeks while they are still nursing. Dogs leave the nest and the care of their mothers at six to ten weeks. During this transition period, they need someone that will respond to their care-soliciting behaviors. If they cannot make that attachment, they almost certainly will starve to death, which, if the niche is full, will be about 96 percent of them.

An appealing feature of dogs is that social bonding usually develops into a lifelong relationship. Because of this, adult dogs will continue to follow humans, sheep, goats, or whatever species were present during the critical period (fig. 12). One finds village dogs the world over sitting among poultry and small livestock without a hint of ever doing any harm to them.

In the United States and other developed countries, dogs are often raised in isolation from species other than humans. This means that as adults they are liable to show their predatory behaviors to these animals and become chicken killers, cat chasers, and sheep chasers.

But the social-bonding feature leads to another phenomenon in the dog world, the livestock-guarding dogs. In Maasai villages, it is quite usual for wee pups to come in from the surroundings and hang out around the houses and cattle, often soliciting care from children. The children feed them and play with them.

They grow up together. The Maasai boys go through intense puberty rites when they are about twelve years old, and after healing,

Figure 12. Most mammals and birds learn what species they are during a critical period shortly after birth. Bonding depends on what animals or objects are present during that critical period. Here we see dogs bonded to chickens, pigs, and goats. (Top and bottom photos by Daniel Stewart; center photo by Isabelle Coppinger.)

it is their turn to go into the field with the cattle. Their dogs go with them (fig. 13).

The boy and the dog both achieve a new status — the boy is now a herder/warrior and the dog is now a livestock-guarding dog. In conversation, an older warrior was asked what the dog was supposed to guard the cattle from, and he said, "Lions!" "Lions? A lion will eat that dog!" "Yes," he replied, "but we hope it will bark first."

Almost all pastoral societies have their dogs. In many, the pups will solicit care from and be bonded to the shepherds, or future shepherds, or perhaps with the livestock themselves. With many sheep, goat, or cattle herders, the industry with which they are involved is the making of cheese. Thus, a small pup sitting in a milking area, whether it is indoors or out, will get the leftovers from the milking and cheese-making process. Throwaway products such as whey are a major component of the diet of all the age groups of these dogs.

An important point here is that the shepherds and the livestock in a sense all smell alike. Among members of the dog family, the bonding process is primarily olfactory. Why didn't the gaggle of tethered dogs at the sheep fair not bite the shepherd? Because he smelled like a shepherd. Why don't the dogs in the Mexico City dump bite the residents? Because they smell like the dump they live in.

In whichever case, the dogs grow up in that livestock community and so they are part of that community. They can be bonded to many different aspects of that animal environment. Depending on the various circumstances, they will follow the shepherd or sheep or goats. We once did a study in Italy's Gran Sasso, a great huge grassland on the edge of the Apennines covered with flocks of sheep, their shepherds, and their dogs. Our research question was, "Were the dogs following the shepherds or the sheep?" The answer, gleaned from several weeks' worth of data sheets, was that about 60 percent of the dogs were following the sheep, 30 percent were following the shepherd, and the other 10 percent were still at the camp where the sheep were getting milked at night. In the latter case, it was hard to know for sure, but it

Figure 13. In Maasai cattle country, little pups come in from the bush and often are adopted by little children. When the boy becomes a man, the pup becomes a livestock-guarding dog.

looked like they were bonded to the milk cans. Improbable as that may seem, reflect back to the great experiments that Konrad Lorenz did imprinting ducks while working in his backyard. Some of the ducks were imprinted on the garden water faucet. After all, the milk cans smell like the shepherd and the sheep, and if that is where the dog's food appears, that is where the dog will wait for human waste.

Most of the dogs would follow the sheep (or shepherd) each day from the milking station into the grassland, though they would not always go with the same flock or the same shepherd. In the Mexico City dump, many of the dogs would follow a person around the dump all day or follow someone home at night. One woman there said to us, "Oh, I have a new dog tonight." The dog had selected her. Why wasn't the dog loyal to one person—"man's best friend"? Because the dog is loyal to a smell. It is the smell the dog was imprinted on, and that's what it is "loyal" to.

For nomads or transhumance peoples, following the sheep around often means a great long walk. We have been on some of these migrations, where the shepherds, sheep, and dogs walk anywhere from 200 to 500 miles in a matter of weeks. In a typical transhumance, the shepherds' wives or family ride horseback on ahead and set up camp for the night and cook the food for the shepherd. They have pastures along the way where the sheep (or goats or cattle) can graze. And what about the dogs? Well, what about the dogs. They do what dogs do, of course—follow the smell.

A long walk across the countryside and into towns and along streets and across highways is dangerous for dogs, and many of them get killed. Quite a few get lost. In one study we made on American ranches, 40 percent of the mortality among the dogs was due to their being lost—lost, that is, to that community but not necessarily dead (fig. 4, *bottom*). The transhumance dogs are hit by cars, and it's not uncommon to see dogs with broken legs continuing on the trek. The migrating shepherds are in charge of caring for and moving the sheep to new grass. But they are not responsible for the care and movement of the dog. Village dogs such as these are bonded to people or livestock, though not because of anything people have done: the dogs follow because that is what dogs do when they grow up in that livestock environment.

In this day and age, transhumanent livestock are moved mostly by truck between winter plains and summer mountain pastures. When the sheep are loaded onto the truck, the dogs can get in with them—or not. Whether they get on the truck with the sheep depends on whether they are bonded with the sheep, the shepherd, or the milk cans. If they don't get on the truck, they might be faced with a village with nobody in it, where for months human wastes are not being discarded. The dogs are going to starve to death. Can they go and scavenge in a nearby village? That depends if there is a nearby village with people in it. It depends on whether the nearby village (niche) is full of dogs or not. It depends on whether the nearby village has the social-bonding signals the dog is accustomed to, such as the smell of sheep or shepherds or milk.

Riding on a truck for the first time might be frightening for dogs, and they may not want to do it again. When the truck stops along the way, and the shepherd lets the sheep out to graze at a roadside, the dogs will get off the truck with the sheep. When the truck is reloaded, it is the dog's responsibility to get back on: remember it does not necessarily have a name and certainly has not been taught to come when called. Many do not get back on the truck. The trail along a migration route will be strewn with lost dogs—dogs that did not get back on the truck. Many of them end up at local dumps. In the years that we bought livestock-guarding dogs, it never dawned on us that we could easily have captured dogs along the migratory path where they were lost. Dogs sat in local dumps, their only fault being they did not get back on the truck.

Whether killed by accident or lost along the way, their fate is due to natural selection. The dogs are lost to the transhumance culture. They are very likely reproductively dead because finding a place in the local niche is problematic. They may not be adapted to this new geography. They certainly may be outcompeted by the local dogs. We will see this again with the discussion of stray dogs, which also do not compete very well with the resident dogs. In addition, local people will often shoot or poison these dogs in dumps, which are now too full of dogs.

In that special relationship between caregivers and dogs, it is as often as not that the dog initiates the relationship. In that special relationship between people and dogs, the dog's survival from neonate to adulthood depends on the dog making that relationship. The dog's reproductive survival depends on initiating a connection where it can eat in the presence of humans. Of course, making a mistake along the way is often lethal.

12

People Adopt Dogs

Biologist E. O. Wilson described the special relationship between humans and animals as an instinctive, innate, and inherited attraction. He termed it "biophilia." Dog people would like to think this special relationship applies especially to dogs. Caring for a dog, they posit, is calming and leads to a longer life. Dogs, they therefore conclude, are good for your health.

The problem is that the argument for biophilia—the need, simply put, to be with another animal—rarely ever takes into account the number of people that seek psychiatric help for animal phobias. Of those suffering such phobias, some 30 percent are cynophobic—afraid of dogs. Nor are the 806,000 Americans per year who had to find a doctor or go to the hospital to have their dog bite treated asked if they had changed their mind about man's best friend.

Asking people in Mexico or Africa why they have dogs often brings the reply that pups are cute and/or they feel sorry for them. Children are different. We found children very much involved in adopting wild things. In East Africa, some young girls (plate 12, *center*) had adopted a little nestling bird. They had a string tied on its leg, and they would let it fly out and haul it back. They would walk it around as if it were on a leash. At the end of the day, the little bird was dead.

A first rule in adopting any species is: the species has to be adoptable. Most wild things are not easily adoptable. The little bird, for instance, was not of an adoptable species. You just cannot take a little bird from the nest and feed it. You have to know something about bird diets, including the food value, the water content, and the digestibility. Nestling birds have tiny little stomachs and a high growth rate, which means they have to be fed continuously from dawn to dark with food that is often partly digested by the parents. Also, most wild things have a specific temperature range they are adapted to, and if it goes much beyond that range they quickly die. Tropical fish tanks, for instance, come with heaters and aerators.

Wolves aren't all that easy to raise and tame, but when you do it correctly they can be handled as well and trained like a dog (plate 15). Dog pups, in contrast, are a cinch to adopt. They may be the easiest of any species to adopt. They can and will gorge feed, which is unusual for infants in the animal world. They can eat almost anything. They can eat solid food, starting in the nursing stage at twenty-five days. They can eat anything a human can and lots that humans cannot. Cooked grain or meat, fecal material, and rotten vegetables are all possibilities—the waste products of the human diet. They have teeth and can tear and chew at an early age. They even starve better and longer than most animals. Also important is that they can tolerate high levels of abuse and handling. A youngster in Mexico who had adopted a dump pup demonstrated to us how well his dog was trained to sit and stay. The boy towed it back and forth along the street on a leash. The pup looked really good and unlike the little bird appeared to have no problems being adopted.

Why do children adopt so many animals, such as birds or dogs?

The answer is that kids around the world are the only members of any society with a surplus of time. Unlike their parents, who have all kinds of projects to keep body and soul fed, little kids around the world often do not have much to do.

Why do children adopt dogs more often than other species? It is not a given that children prefer dogs. There just happen to be more pups available around a village for adoption. Not only are tiny pups more available than most other species at any time of year, but it may just be that more dogs survive the adoption process than other species. It may also be that dogs are more interesting to adopt. They solicit care from people. They actively beg for food and will do anything to get it. They are, therefore, the perfect pet, and their lives depend on somebody paying attention to them.

Another reason for the popularity of dogs with children is that pups can be, and often are, sold to tourists, allowing the kids make a little money, which is otherwise not usually easy for them. In the mountains of Venezuela, kids stood along the roadside holding small pups out to passing cars like ours. The area had a reputation for being the birthplace of Simón Bolívar's famous war dog, Nevado.

Kids capitalized on the idea that these pups were direct descendants of Nevado. We had similar experiences in Mexico, South Africa, Macedonia, and even Poland. When anyone showed any interest in the local dogs, a ten-year-old would show up with a pup for sale. We certainly fell for a pup now and then.

13

People Breed Special Dogs

Generally speaking, when people think of dogs in developed countries, they think of people as having purebreds, along with a few "mistakes" of interbreeding, commonly known as mongrels. Back when we were kids, there were all kinds of stories about how dogs were transformed from wolves or jackals directly into breeds. There were even books that claimed big dogs came from Chinese wolves—the big wolves of Mongolia. In the 1950s, Konrad Lorenz toyed with the idea that some breeds descended from the wolf and that most other breeds were descendants of jackals. The problem is that once an idea gets into print, it stays in people's perception for years. The idea becomes a factoid.

The 850 million dogs in the world that are not under human reproductive control give the impression that humans had little or no influence on what these millions of dogs look like or how they behave. That probably is not quite true, however. In various places in the world, the village dogs vary in unexpected ways from the basic thirty-pound model, including in shape and color. Some of that variation — or lack of variation — can be attributed to human causes. As we traveled, we would find dogs in a particular region or even in a village that are distinctive. What we want to know is: Have humans in some way contributed to those variations?

So far, we have discussed the weight of all the (village) dogs in the world and found they all weigh in the neighborhood of thirty pounds. From this, we concluded that size is an adaptive characteristic and a product of natural selection.

Sometimes, on inspection, one finds that the distinctive characteristic is an adaptation to a specific environment. It could be a peculiarity of climate or a peculiarity of the dogs' niche. For example, the dogs accompanying nomadic or seminomadic pastoralists in the mountain regions of Eurasia are twice the size, bulkier, and rougher-coated than the smaller, smooth-coated, skinny dogs of equatorial shepherds. Size sometimes seems related to how far one of these nomadic dogs has to walk and in what kind of conditions. Not only have they commonly had to walk several hundred miles twice a year, but they have had to be gaited to keep up with sheep or cattle or whatever livestock they are bonded with. They have, in effect, evolved a walking shape. Exaggerate this discussion and think of the extremes, with maybe a Chihuahua-sized dog trying to walk several hundred miles along steep mountain trails. The poor little thing would have to take so many steps it might come apart. Could the Chihuahua have enough stored energy to walk all day long, day after day? Such migrations would select for bigger dogs with an easy, ground-covering, efficient gait.

A bigger dog does not mean, say, a Saint Bernard. Having an animal the size of a big Saint Bernard try to walk that far, especially during hot weather, would be cruel. It is our benefit/cost problem again. The Saint Bernard has to carry its own 150 pounds — just try to pick up and carry 150 pounds all day long. A dog of the migrations has to have a long gait and at the same time not be too big. On one transhumance migration that we followed for a bit from the lowlands of Greece into the Šar Mountains bordering Albania, Macedonia, and Serbia, most of the dogs were in the sixty-pound range or, in some cases, maybe a tad bigger (fig. 14, *top*).

Big, chunky, and furry dogs have a selective advantage in a cold country and are at a selective disadvantage in a warm country. We had an unforeseen experience when we introduced a small population of dogs from Turkey to America and then, after a few years, into Namibia, Botswana, and South Africa. We had picked big dogs for the Namibia project, thinking they needed to guard the stock against lions as well as cheetahs. The problem with those big dogs was they could not stand the heat on desert scrub land. They panted a lot. Some of them ended up with sunburned tongues, and like many sunburned tissues, these developed into cancers.

Being in a climate that is too hot reduces the average longevity of a dog and, ultimately, its reproductive success. That is what natural selection is all about. Cheetah Conservation Botswana ended up having great success with the thirty-pound native pastoral dogs. The only problem with the thirty-pound livestock-guarding dogs there in Botswana and in South Africa was they were unacceptable to the white ranchers, who were convinced they needed a specialized breed of dog, which had done the job for thousands of years in some faraway country and was the product of careful breeding. Like many of us, they were convinced that show breeds would perform the livestock-guarding behaviors for which their ancestors had been selected.

The dogs we brought in were exactly that—well, sort of. Though they had been carefully bred for maybe ten years in America, some of them rolled over and died as soon as they got off the plane in Namibia: they caught all the local tick fevers and had terrible troubles with spear grass. These were physical complications that the "local pavement specials" had little problem with.

Regional variations in dog populations are called landraces. Many

Figure 14. The Tratturo is a historic path of a transhumance migration. Anciently, these were long walks by sheep in the spring from winter pastures to summer pastures and then back in the fall. Millions of sheep make these long migrations every year with their shepherds and dogs.

regional characteristics are just straight old Darwinian natural selection: selection for viable shape, size, and activity patterns for the specific region. What is not always immediately obvious is that the selective force—such as the semiannual transhumance migrations—might only happen for a couple of weeks a year. Something that causes extreme stress for a very short time and affects the population is called a bottleneck. The pastoral dogs have to be able to survive for a couple of weeks twice a year on the long walk.

As we mentioned before, sometimes the dogs just drop out of the walk or get lost along the way. They might not be dead but they are lost to that culture. Selection is for the dogs that make it all the way, and then all the way back. When we go to the high mountains and find some unusual dogs, it is not always obvious that they are the result of having survived a long walk.

In some regions, the dogs can vary in a characteristic such as color, so that, say, every dog seems to be a different color and/or color pattern. In that case, the biologist might conclude that there were no selective forces working on color—any color will do. Animal ecologist Alan Beck once suggested that the variation in colors and patterns made individual dogs distinctive (an oddity effect) and therefore more attractive to humans. If that were the case, then the variation, where every animal is different, is a selective advantage to individual dogs. Small pups soliciting food from villagers might make out better if they had a distinctive color. Thus the adult population with a variation in colors and patterns may have been selected for. That is, what is being selected for is continuous variation in the population.

Quite often, when one goes into a mountain region, say in central Italy, almost all the pastoral dogs are white. At least three hypotheses come to mind:

1. Natural selection favors white dogs in this region.
2. The white dogs are a consequence of a founder effect.
3. People are selecting for white dogs.

Natural Selection

Take the first hypothesis: natural selection favors those mountain dogs with white coats. That means that white dogs have a greater chance of reproducing than dogs of other colors. Many species of mammals and birds have white coats, such as polar bears, ptarmigans, and snowshoe hares. There are also white wolves. Most of the time, species that have white coats are associated with hunting or hiding in the snow—a cryptic coloration.

That does not seem likely for these Italian dogs because they are only in the mountains in the summer, when there is no snow. In the winter, they are on the grassy plains along the Mediterranean, where there also is no snow. One Italian veterinarian explained to us that the sheepdogs were white because the wolf would not see the dog in the flock of white sheep, giving the dog more of a surprise advantage over a marauding wolf. When it was pointed out to him that most sheepdogs of the world were not white he said, "That is why the Italian dog is the best."

As biologists testing the theory that the white dogs are best, we might do the experiment of having a control group of dogs of other colors tested against a group of white dogs and see which survived longer and reproduced more often. We analyzed data from the 1,500 dogs in our livestock-guarding dog project and never found the white breeds surviving any better than the breeds of other colors. Some slight difference in success did occur. Many of the white breeds tend toward pink skin, noses, and occasionally pink eyelids. Lightly pigmented dogs were much more susceptible to sunburn than were those with good black noses and eyelids. When we were racing sled dogs, we tended to avoid white dogs because they had softer feet that damaged more easily.

Besides, as already suggested, wolves do not tell the difference between sheep and dogs by color but, rather, by smell. Also, no sheep that graze in these pastures are white except during an agricultural fair when they have just been washed. Sheep are a dirty gray color, and we could always see the white dogs from a mile away.

If one looks at color frequency of village dogs around the world, it seems to us that lion-colored dogs exist at the highest frequency. That is just a guess, but if it were true it might make one suspicious that the dark tan is an adaptive color. Why are all lions the same color? It must be adaptive.

Founder Effect

Our second hypothesis was that the white dogs of Italy might be an example of a founder effect. When a small number of individual animals are the initial population in an isolated region, they are known as the "founders." We discussed the founder effect in some detail in chapter 4. But it is such an important concept in understanding breeds of dogs that a brief review is warranted.

If the Italian shepherd dogs in the Abruzzi Mountains were the result of a few introduced white dogs in the isolated region, then the resultant population would show a high frequency of white dogs. That hypothesis is reasonably difficult to test.

It is true that the number of dogs in local populations of dogs rises and falls fairly regularly. The variation in numbers is often the result of lethal diseases. After the pestilence, the population will return to their former numbers, whatever the niche will support. However, they will do that with less genetic diversity. The new population will be represented by the genes of the few founders and not the variations that existed in the former population.

Let's imagine that we have a population of multicolored dogs, ranging from black to white, with all the colors and patterns in between, and that every color and pattern exists at equal frequencies — meaning: there are as many black dogs as there are white dogs. Now all the dogs in that population — say, the Italian mountains — get this terrible disease, and all the dogs die except one lucky pregnant, white female. Alternatively, one could imagine they all died and after a dog-less period, a pregnant white female strays, by chance alone, into the region from a neighboring population. From that single, pregnant female the niche (our bowl of marbles) fills up to full again. The frequency of white dogs in the recovering population will be much higher, and indeed the new population could be all white if she had just happened to mate with another white dog.

The founding individuals (all offspring of the pregnant white female) of the new population cannot possibly have the genetic diversity for all the colors that existed in the prior, multicolored population. When you go to regions where dogs show an unusual character, it is reasonable to think that this is the result of a founder effect, particularly if certain coat colors are prevalent in a large percentage of the dogs. It is possible that the Maremmas of central Italy are white because of a founder effect.

Selection for White Dogs

The reason of why neither natural selection nor founder effect is probable in the case of white sheepdogs in Italy is something we have not yet mentioned. The shepherds here, for whatever reason, like white dogs. They think a sheepdog must be white. They think the original sheepdogs were white. They think that if you do not have white sheepdogs you are not a good shepherd. They think that if a dog is not white it is a mongrel—in Italian a "bastardo."

Now, when you read the breed literature, it will say that Maremmas are white and that these Italian pastoralists have bred their dogs to be white since the beginning of the Enlightenment. Thus, the conclusion we came to when we started our studies on these dogs was that the Italians were breeding white dogs to white dogs. People who imported white Maremmas to the United States liked the white dogs, as did breeders who imported other sheepdogs from the mountains of France (Great Pyrenees), Turkey (Anatolian shepherd), and the former Yugoslavia (Šarplaninac). We thought that Maremmas were a purpose-bred breed and that Anatolians were a breed and that the Šarplaninac was a breed.

After years of study, however, it became obvious that the transhumance people were not breeding dogs at all. It is nearly impossible to believe that these seminomadic pastoralists could sexually isolate a population of dogs. There they are on a faraway mountainside with their sheep and dogs, and their favorite female comes into heat. How do they isolate her from other males? It would be impossible: those dogs might not even have names, the shepherd might be a hired hand, and the dogs are not with him but rather with the sheep. What happens in those situations is that the male dogs, moving with the flock, have first access to a soliciting female. They may all breed her. There can be as many fathers as there are pups. If she is white and all the male dogs attending the flock are white, then all the pups are going to be white. It may be difficult (but not impossible) for an outside male, meaning an unfamiliar male, to gain access to the female. Whether outside males, perhaps males with spots or some other coloration, can gain access to her depends a lot on circumstances. If the flock is on migration and traveling the roads from winter to summer grasslands, the likelihood of an outside breeding increases.

In one of our field studies in the Gran Sasso d'Italia mountains, the shepherds told us they had decided to try a herding dog to work their flocks. They acquired a German shepherd dog, and the next generation of Maremmas showed he had been sexually active. Nevertheless, because there can be more than one father to a litter, not all the pups in any one litter would be hybrid German shepherds. Many would be white.

So how do they maintain white purity? Once, when asked why all the sheepdogs in the Molise region of Italy were white, a milker just laughed and said: "Because only the white ones are allowed to live." In other words, they cull nonwhite pups. The frequency of white dogs is high in the region, thus the frequency of white pups is high and is maintained by eliminating other colors.

A slight digression: one time, when we were searching Turkey for the dark karabash Anatolians, we spotted a beautiful male in a village as we went by. We went in to see if we could buy the dog. No, it was the shepherd's favorite dog, and he needed it. "Would you like to see the pups from this dog?" he asked, and of course we would. Two beautiful male pups about five weeks old were presented, and we oohed and aahed over them and offered to buy them. No, he told us, he needed them as replacements. "Where," we asked, "are the female pups?" "We threw them away," he responded. There it was again: culling the unwanted pups. In actuality, however, because they felt they could not kill a dog because all living things belonged to God, the story has a happy ending: we got the shepherd's uncle to show us where the pups were thrown. A water buffalo had stepped on one, but the other two were in pretty good shape. We found a ewe with milk and let the pups nurse. We took them home, and Peaches and Cream became not only great livestock-guarding dogs but also founding stock for the Anatolian Shepherd Dog Club.

Call it dog breeding if you want. Maintaining a regional color pattern is a reasonably standard procedure throughout pastoral cultures. The color becomes the signature for a good sheepdog. By the process of natural selection, the dogs perform the long migration, avoiding hazards as much as they can along the way. Their population numbers are limited by food availability and whatever else are limiting variables. They are socially bonded to shepherds, sheep, or milk cans. They are selectively culled at birth for what Darwin called "capricious characteristics," such as color. Like all free-breeding populations, they produce too many pups, and the culling does not change the total population size.

Humans can affect the phenotype (the way the animal looks, for example, size, shape, color) of dog populations in other ways as well, making those populations look like they are products of careful breeding. The most common and unintentional human behaviors that create breeds are favoritism and differential support. The pup that is the cutest, whitest, or biggest is attractive to a food-giving human. The favored individual is more likely to survive than its littermates, and survival means being able to reproduce and pass its capricious characteristics on to the next generation. Such characteristics must not have deleterious effects that limit the animal's performance. We mentioned above that some livestock-guarding breeds are too big for some habitats and badly overheat, while some white breeds will not have enough pigment around the eyes and nose, which will end up sunburned on treeless pastures.

We have described three methods of producing animals that look like a breed without any intention of creating a breed. Thus, we have what look like breeds of dogs for which there have been no "arranged" matings performed.

Favoring and supporting certain animals or culling individuals that are not wanted is called postzygotic selection. The zygote is the egg. The egg results in a pup that is supported or not and culled or not. The pup arrives and is on the ground and then somebody makes the choice. These are not breeds in the modern sense, and a population geneticist would call them landrace dogs. In a wild species, they are called subspecies. A subspecies is defined as a geographically based population within a species whose genes show a nonrandom distribution of alleles.

In our culture, we tend to practice prezygotic selection. People pick and choose the breeding pair. The expectation is that dogs are the products of sexually isolated breeding pairs. But there is a halfway method. Dog breeders in isolated communities might not have control over mating but they can have a favorite female. Thus the breeder might not know who the father(s) was but has carefully cared for the mother and her pups.

In antiquity, records were often of the matriarchal genealogy. It is said, for instance, that in Alexander the Great's celebrated horse-breeding programs in the Middle East the lineage was recorded mother to mother. Good breeding of hunting dogs in Tanzania was created "through the mother." It is not exactly postzygotic, and it is not exactly prezygotic, but it works.

Prezygotic selection indicates that the breeder picks and purposefully mates the parents. Making a breed quite often started with people capturing a few of the landrace dogs and sexually isolating them. A visitor coming from far away to a beautiful mountainous watershed where the dogs are white might think something like: "These people have wonderful white livestock-guarding dogs. These are an ancient pastoral people and their dogs are white — the original sheepdog must have been white. These people are breeding white dogs."

Since about the 1930s, people have been capturing dogs in pastoral societies, starting with the Pyrenean mountain dog, and taking them to their home country in the West. There they sexually isolate them from all other dogs and develop a breed that can be registered by the national dog breed registry.

We searched throughout pastoral societies for working livestock-guarding dogs. When we were in Turkey, we had set out to find reddish-tan dogs with black faces because that was what our sponsors wanted. As young travelers, we were led to believe that true Anatolians were these karabash (again, meaning "black head"), and we thought at the time that this is what the Turks were breeding. Interestingly and unknown to us, other travelers thought the real and original Turkish dog was white — an akbash (again, "white head"). In order write a "standard" for the breed to be accepted by breed organizations such as the AKC and the Kennel Club of Great Britain, quite a few tetchy communications ensued among breed fanciers about the "original" color. At present, arguments abound about which region within Turkey produced — 6,000 years ago — the prototype dog of which color.

What most breed organizations don't seem to realize is that the few dogs they capture and export do not and cannot have the genetic diversity of the population in which they originated — because of founder effects. Starting a breed from a few founding individuals soon leads to inbred strains of dogs, which ultimately have less genetic variability than the founding stock. The golden retriever breed traces back to two founding males — and they were brothers. The Border collies trace back to one individual, and the Siberian huskies (another "ancient" breed) trace back to a handful of dogs at a kennel in New Hampshire. The basenji, another of the mythological ancient breeds, was registered with the AKC in the early 1930s from nine individuals brought to the United States from the Congo. It is a familiar and disturbing story. Is there any "breed" in the late nineteenth- or twentieth-century United States that started from more than a handful of captured landrace dogs, and not further back than the nineteenth century?

A geneticist colleague of ours at Cornell University, working with breeds of cows, did the math and determined that if you start a breeding program (meaning closing the gene pool — sexually isolating your stock) with 500 unrelated males, and you allow them to breed at random with females, inbreeding would start by the twelfth generation. Thus, if you start with a handful of males and females that may or may not be unrelated and do not let them breed randomly, in which future generation does inbreeding start? It might be as soon as the first.

At the end of the nineteenth century, our modern definition of breeds began with the concept of eugenics. Sir Francis Galton coined the word "eugenics" in 1883. The word, created from Latin and French, simply means "wellborn." Eugenics is a system of genetically directing a population through nonrandom mating to achieve a perceived standard.

The Europeans started eugenic dog-breeding programs through the newly emerging kennel clubs. The concept we have now of prezygotic breeding, meaning planned mating within a closed gene pool, started in that post-Darwinian era. There were attempts, at that time and following well into the twentieth century, to create well-bred people as well as well-bred dogs and other domestic animals.

Eugenic mating of dogs — that is, improving and maintaining purebreds—will eventually fail for much the same reason it did in humans. Two basic problems arise quickly. The first problem, always, has to do with who is choosing the qualities to be enhanced. With the beginning of modern dog shows and the judging of visible characteristics, subjective decisions caused major dissention in the field. Blue ribbons were awarded for traits Darwin would have defined as capricious. It is reasonably ludicrous to think that a judge could look at a golden retriever and tell you its value as a hunting breed or if it could perform some task or represent some ancestry based on whether it has a perfect golden retriever shape.

The second difficulty is that eugenic or prezygotic breed selection is designed to decrease genetic variability. In the end, breeders don't want many different colors—they want one color. They do not want variation in size; they want the "perfect" size. The idea is that they want the breed to be a narrow representation of what they think the original golden retriever or the original Jack Russell terrier looked like. The goal is to preserve that image and only that image—in spite of the fact that journal papers have shown that breeds such as English bulldogs or Saint Bernards do not have the head shape of their working ancestors of a hundred years ago.

Eugenics, as we have just said, is designed to decrease genetic variability. It truly does just that, and it does so on purpose. The dog that has won the best-of-breed ribbon in the show is later selected to mate to preferred females. The next generation will have more of the champion's genes and fewer genes from all those other dogs that were not judged to be of top breed quality. Those inferior dogs varied from the mean. Thus the breed suffers the purposeful loss of genetic variability. It is an extreme founder effect. The loss of genetic variability by nonrandom mating is unwise when applied to people, and some of us think the same is true for dogs. In the end, this system can have dire consequences for the physical and mental well-being of those "purebred" dogs or people.

Often the characteristics that are being selected for are not exactly genetic. At the beginning of the eugenics movement, the idea took hold that the reason some people were rich was that they were genetically better—they were of better "breeding" stock. In many cultures, royalty could only breed to royalty.

Many dog breeds started because the breed's founding population could be trained to do some task not only better than any other dogs but better than any other species as well. Golden retrievers are not the best hunting dogs because they are golden colored but rather because the Scottish businessman Dudley Marjoribanks (Lord Tweedmouth) bred good retrievers to good retrievers, regardless of their background or breed, wherever he could find them. He put years into training them properly, but he also had this little quirk: he preferred that golden color. Like the mountain shepherds, he got that color by culling or selling the nongolden colors postzygotically. Ultimately, though, what was most important to him was achieving superior hunting ability, and he would breed his dogs to any dog that was a superior hunter anywhere on the British Isles—regardless of its size or color or ear carriage or any other capricious trait.

Breeds can be created in many ways, including starting with a land-race dog or starting by crossing existing breeds together, then closing the studbook and only breeding dogs selected for certain traits. A skillful breeder can have, within a few generations, an identifiable new breed. Hundreds of examples can be given of this method, including some of our most popular breeds, such as the German shepherd, the Doberman, the Rottweiler, and all of the coursing breeds.

When the studbook is closed, the dogs in it supply an initial founder effect. Because those few animals chosen as the founding stock cannot genetically represent the population from which they were derived, genetic variation has been cut off. In the following generations, nonrandomly breeding the best to the best results in further loss of variation. Eventually every "breed" treated this way will get into what is commonly called inbreeding depression.

Frankly, we think it is surprising that breeds last as long as they do.

Dog fanciers have believed that the best cure for genetic problems is not to breed to dogs that show the problem. For years, dogs were tested for hip dysplasia, for instance, and dogs with bad hips were not bred. But that further decreases the genetic variability. Regardless of whether it gets rid of the problem, it actually puts the breed in more jeopardy. When problems increase, breed clubs will, in desperation, introduce individuals from outside the studbook. At that point, however, bringing in a few "outside" breeders from another gene pool is futile. Usually when breed clubs have opened up the gene pool to other breeds, they specify which breeds. Hybrid vigor is the underlying theory. Though this method has been used in many livestock programs, it is only effective for one generation. The first generation will be uniform and probably healthy, but breeding to those individuals increases the variation not only in the gene structure but also in the phenotype, and the process of bringing them to a created, described standard starts the problems all over again. It is futile.

Allowing purebreds to breed randomly within the sexually isolated population would be better. And even better than that would be to let a female breed with many males, producing litters with many sires and culling those pups that don't meet one's standard, as the pre-eugenic breeders of hunting and working dogs did.

14

Breed Genes Stray into the Village Dog Population

With over 300 breeds of dogs and 75 million pet dogs in the United States, it is not surprising that some of them leave their families and yards and go native, straying back into the village dog community.

Why do they do that? Being in the village dog community with other dogs, whether in town or a dump, may simply be a more rewarding life for a dog. Dog owners really have to work to prevent them from straying back into that larger dog population. Being shut in all day and taken for a walk on a leash once or twice and eating the same brown food every day, year in and year out, neutered, and often disgustingly overweight might not be a very comfortable life.

Many people think dogs have special cognitive skills that other mammals — even wolves — don't have. If that were true, life in the house would make them even more bored and worse off in their daily lives. We began our latest study of dogs in natural environments like dumps because they were a more fascinating animal than those in our homes.

We are just joking, of course. We doubt that dogs can be bored or consciously "stray away," looking for a better life. They don't have that kind of brain. In our experience, most of the pet "dogs of breeding" that end up independent and unrestricted in the streets or dumps are just lost. The transhumance livestock dogs that don't get back on the truck are lost, and they often stray into a nearby dump: that's where the food is. In the United States, police stations keep a lost-and-found record just for dogs. Sometimes, when people are on a trip, the dog gets out of the car when nobody is looking. Or in our case—we live on a tree farm and a spur of the Appalachian hiking trail follows along our rural road—occasionally one of our yard dogs will follow a hiker and, perhaps, their dog(s), for miles and become lost. Luckily, we have a local radio station that will put out the message—"the Coppingers have lost another dog"—and we'll get it back. One time a women said, "Oh, I'd hoped you wouldn't have found Shak, and we could have kept him. He is such a nice dog."

Last week, from ten until after eleven in the evening, a houndlike bark far out in the forest made its way slowly from northwest of our house to southwest. The next afternoon, out in the field between the house and the trees, we spotted a stealthy little beagle eyeing a very nervous turkey that held its head stretched as high as it could, watching the dog. Luckily, the dog was amenable to being carried to the house, where its collar revealed the owners' telephone number. It turned out the owners had been searching and calling throughout the forest for two weeks to try to locate their dog. They live about two crow-route, forest-filled miles from us; they were teary-eyed grateful to have their (thirteen-year-old!) dog back. They said Maggie looked a bit thinner. She apparently didn't know it, but she had been a lost dog.

We first came into considerable contact with real stray dogs in the Tijuana dump. They were noteworthy because a large number of them were pit bulls. Dumps are not often in the most expensive parts of town and that was true of Tijuana. Outside of the dump were row houses — family houses. If you lived in one and managed to improve your earnings, you could move to a different section of town. Day laboring seems a common way of making money in Tijuana but not big money. In Tijuana, dog fighting could mean big bucks. The audience often came across the border from the United States, where dog fighting is illegal. They would bring big money to bet on dog fights, making that look like an easy way to make a lot of money. Buy a pit bull, and all you would have to do is win one fight. One fight with the right odds, and you have a windfall.

The problem, of course, is that most people, including those in the row houses in the neighborhood of the Tijuana dump, don't know how to train a fighting dog. People often think fighting is genetic, and all you have to do is get a well-bred dog and it will fight — the idea being that a well-bred pit bull could earn you a fortune. If the pit bull is kept at home, the dog often becomes a nice dog and fun for the kids to "adopt" and play with. We'd find them dressed in fancy clothes, part of little kids' games.

Most of those dogs never see the pit. Most of them end up as the family dog, partly dependent and mostly unrestricted. They are rarely neutered because people either don't think about such things, or they don't have access or money for the procedure, or they think there might be some money in breeding. Since the dump is close by, these dogs spend their time as day migrants, socializing with other day-migrant dogs in the dump.

Their dump behaviors are quite different from the dump's resident dogs, and of course they are a very different conformation so it is easy to tell them apart. Probably if you just saw how they behave, you would quickly notice they are different from the dump dogs. For one thing, they are not as nice in the dump as the village dump dogs — they tend to be more belligerent. Some of us found them a little frightening. They noticed people approach and instead of moving shyly away they would stand their ground. They argue more and spend more time growling and carrying on among themselves. They are often fed at home and so do not spend their time feeding, resting, or even attaching themselves to someone. Dump dogs are always tending to the business at hand, whereas the migrants are more interested in the other dogs, for whatever reason. Often the morsel of food in the dump is of better quality than what they are getting at home and therefore worth defending.

Why is it that they tend to be more aggressive—because they are pit bulls? Maybe that's part of the story. Yet one cannot rule out that they were raised differently. Their early pup behavior developed in a solitary environment, often alone in a house or pen. Then, after weaning, they lived alone in a pen for much of the day, separated from their people, who were at work or school. They were the only dog and because they were valuable pit bulls whose family fortune was at stake, they were locked up somewhere safe from thieves. They did not have to find their food in the dump or even solicit it from some helpful person.

From the point of view of the World Health Organization, these pit bulls in the dump probably started as family dogs, totally dependent and fully restricted. Then they graduated to neighborhood dogs, semi-dependent and mostly unrestricted, and perhaps finally to feral dogs, independent and unrestricted.

Tijuana is unusual in that you can actually distinguish recognizable breeds. Also, we could see that the stray dogs in that population are commonly the result of irresponsible ownership. People acquire dogs and do not take care of them after the initial hopeful period of windfall profits. Therefore, the dog ends up as a stray. This situation at the dump in Tijuana is not typical of the dump in Mexico City, for example.

The Mexico City dump is surrounded by one of the largest cities in the world, with a human population that loves purebred dogs. They promenade them daily; you rarely see a stray. Once we saw a dog that looked very much like a Rottweiler roaming the dump, looking lost. Another time we saw a fox terrier look-alike. However, the 700 dogs in this dump are mainly a uniform group, and to some of us they do not look like either strays or crossbreeds.

Identifying strays is not simple, though. Many of the dump workers live in the surrounding community, and some of them will have a bona fide purebred dog that follows them to work on any given day. On occasion, though, we would see a dog on its own at the dump that had the standard shape of some breed, but it seemed to be a stray. Thus, there are at least three categories of strays: the real strays living in the dump, the unrestricted day migrants, and those dogs that have followed an owner to the dump. Whatever the category, if they are reproductively intact they can spread genes into the dump population (plate 13).

In many dumps, a real stray dog never seems to last long. In most dumps or on most streets, we do not see a lot of evidence of hybrids in the next generation. A pit bull is a big animal in comparison to the village dump dog, and you would think that in the hybrid you would see something that would remind you of a pit bull.

Nevertheless, a leakage of genes from these stray dogs into the village dog population is possible. If the Rottweiler (plates 12 and 13) living in the dump, along with six other village dog males, bred a village female dog, what are his chances of getting a pup into the next generation? What are the chances that his offspring will be one of that 4 percent that gets to grow up?

If the theory is right that our village scavenger is like any other animal species competing for limited resources and that its size and shape are adaptive, then one could imagine that the strays and the hybrids would be selected against. Remember that juveniles may have only a 4 percent survival rate, which means competition in the Darwinian sense is fierce. Any dog that does not have the right size and shape, which is very probable for a hybrid—a mongrel—might have less of a chance.

Every village and every dump is intriguingly different. One time we took an animal behavior class to the village of Mucuchíes in the mountains of Venezuela to look for the birthplace of Simón Bolívar's favorite dog, Nevado. It was a beautiful village of about 2,000 people. The waste from 2,000 people theoretically should support 140 dogs. We counted as best we could and concluded there were at least 200 dogs.

As Luigi Boitani and his colleagues would predict, the dog population included family dogs, some of which were recognizable breeds. We also saw neighborhood dogs, most of which looked like village dogs anywhere in the world, except they were a little lighter in color than one might expect (fig. 10, *bottom*). Then there was another population in town that was atypical in that they were bigger than anticipated. Not only were they bigger but they looked like crosses with the Saint Bernard. A third population of feral dogs came into the town at night but we never studied them. They were shy of people.

In the middle of the town was an obelisk with the bust of Simón Bolívar at the top (fig. 15). On the pedestal was the statue of an Indian lad, Tinjacá, along with Bolívar's famous war dog Nevado. Tinjacá and Nevado were natives of Mucuchíes.

The story is that Bolívar (known as the Great Liberator) was returning from a battle and passing through Mucuchíes when a big dog appeared and threatened him and his soldiers. The soldiers were going to kill the dog but Bolívar (a dog man) liked the dog. Bolívar was called "old iron ass" because he spent so much time in the saddle his calloused buttocks would wear his pants out from the inside. He loved horses and dogs and, like General Custer, took his hunt along with him everywhere. Bolívar (as well as Custer, Napoléon, and Wellington), in the tradition of great generals, hunted with his dogs before battle.

On that long-ago night in Mucuchíes, the dog's owner appeared in order to find out what the disturbance was, and after a conversation with Bolívar about dogs, the owner gave the dog to Bolívar as a present. Important to our story here is the owner also "gave" to Bolívar the Indian lad Tinjacá, who had raised the dog. Slavery was still common in Venezuela at the beginning of the nineteenth century, and Tinjacá was now Bolívar's slave. According to one story, Bolívar would say to Tinjacá, "Call the dog and if he doesn't come I'll punish you severely."

Figure 15. In the Andean village of Mucuchíes is a memorial commemorating Simón Bolívar's war dog, Nevado, and dog handler, Tinjacá. *Inset*, Saint Bernards are allowed to run in the streets in an effort to restore the village dogs back to their original grandeur. (Photos by Gail Langeloh.)

Bolívar, his dog handler slave, and Nevado went to many battles together, and at the end of this story, Bolívar finds Tinjacá mortally wounded. The great general yells at the lad, "Where is Nevado?" Tinjacá, with tears in his eyes, says the dog has been killed, whereupon Bolívar runs to where the dead dog is and, burying his face in the fur, sobs at his loss. So much for Tinjacá.

In a town neighboring Mucuchíes, we stayed in a hotel called Los Conquistadores. The irony is that history shows Nevado to have been one of many well-trained and well-bred war dogs, which were popular with the Spanish conquistadors, who used them to terrorize native Americans. Vasco Núñez de Balboa had one that came from the famous war dog kennel of Ponce de León, who not only famously looked for the fountain of youth in Florida, but also, less famously, had one of the best kennels of Indian-eating dogs in the Caribbean.

The claim about the Mucuchíes dogs, though, was that they were a Venezuelan breed descended from European sheepdogs, such as the Pyrenean mountain dog and the Italian Maremma. With careful breeding and crossbreeding, these sheepdogs had become the national dog of Venezuela.

The people of Mucuchíes we talked with, however, thought that in the beginning Nevado was a neighborhood dog. The streets of the town were most probably the source of the famous breed, the national dog, and the descendants of Nevado and his compatriots. And now those village dogs, over which people had no reproductive control, had evolved into thirty-pound mutts.

When we were there in 1989, the local dogs looked almost like any other village dog in the world, except for a few Saint Bernard types running around independently and unrestricted. It turns out that the local people had decided they needed to do something to restore the breed, so they had released Saint Bernard dogs to roam the streets and breed with their village dogs, thus restoring them to the former grandeur of Nevado (fig. 15, *inset*).

The idea of releasing other breeds into the population to keep the local dogs more robust started with Bolívar himself, who brought a number of big white dogs from France in memory of Nevado and the beautiful dogs of Mucuchíes. Each time it has been done, in the following few years, there would be a couple of dogs looking like hybrids but very rapidly "degenerating" back to lovely little village dogs.

Evidently, when Hugo Chávez became president of Venezuela in 1999, he brought a new national spirit to the country, inspiring another attempt to revive the Mucuchíes breed, which continues into the present. This time, however, it involves breed clubs with fully dependent and fully restricted dogs in kennels. What they want is a dog that looks like Nevado but they probably don't want a breed that behaves like the famous war dog.

The Mucuchíes story is one example of how stray dog genes work back into the village dog population. The story really combines two themes: first, the creation of a breed and then, second, the straying back of breeds into the village population.

From those isolated gene pools called breeds, a trickle of genes seeps back into the village dog population. The stray dog problem is going to turn up most severely where populations of dependent and restricted dogs exist. That doesn't occur in very many areas of the world.

At one time, there was a popular myth that if you took all the dogs that stray from the realm of the purebred and bred them all together, the resulting mongrels would look like . . . the Mexico City dump dogs or the village dogs of the world.

We do not think the few purebred dogs that become sexually active with the ubiquitous village dog populations ever change anything. In places like Mexico City, where you find a large number of people with purebreds that are dependent and restricted, you still do not find that the straying pets have much effect on the breeding population in the dump. Most of the dump dogs look like dump dogs.

15

Dog Genes Stray Back into the Wild

Dogs do not usually stray into the wilderness. Their genes, however, might stray back into the wilderness. Indeed, many wildlife biologists worry about dog genes corrupting, even endangering, members of the genus *Canis*. "Corrupting" here refers to the fact that, when dogs and the wild species breed together, there will be a reduction in the natural adaptive traits of the wild type. When two "species" interbreed, the one with the larger population can swamp the smaller numbers.

Conservationists worry that species such as the red wolf or the Ethiopian wolf could become so corrupted with dog genes that they would essentially go extinct.

Dogs really cannot stray back into the wilderness. They would be out of their niche — out of the niche to which they are beautifully adapted. The wilderness has many niches, but dogs, as complete organisms, are not adapted to those niches, as discussed in chapter 4. They could not survive in the wilderness, especially if the wild niche was already full of a wild species. Wolves, coyotes, jackals, or dingoes already occupy the wilderness niches, and they are very well adapted to them. Dogs would not stand a chance competing against them. Nor would hybrids of dogs and wild types be able to compete with their parent species.

In some of the great old wolf books, such as Young and Goldman's *The Wolves of North America*, are reports about a dog running with wolves, but such reports are mainly stories that somebody reported to those biologists. They are the kind of stories that start, "a friend of my sister's husband had a friend who saw what he thought might be a dog running with the wolves." However, never at a conference of wolf biologists have we heard a single story of a dog running with a wolf pack. Could it happen? Of course, but it isn't very likely

As biologists and conservationists, we are interested in whether dog genes are being admixed into the wild critters. Many feel that the Ethiopian wolf is threatened by domestic dogs in three ways. First, the wolf is partly a scavenger, and when the nomads with their cattle move into wolf habitat with their dogs, the dogs are better scavengers than the wolves. For the short time the nomads are present, the nomads' dogs outcompete the wolves for some part of their diet. The wolves go through an energy bottleneck, which may affect their survival. They may lose some of their robustness and suffer reduced survival skills.

A second threat to the Ethiopian wolves is that dogs and wolves are so closely related they share many of the same diseases. When rabies breaks out in dogs, it can spread quickly to neighboring wolves. A few years ago, when rabies broke out in the Bale Mountains of Ethiopia, thirty of the 300 wolves died. When parvovirus began to kill wolf pups on Isle Royale and in Yellowstone National Park, many blamed the presence of dogs for the introduction of the disease.

Some scientists believe that the wolves are getting their diseases from the same background sources that the dogs get their diseases. The larger the population of canids, the more likely a disease will spread. In Ethiopia, dogs now outnumber wolves by a thousand to one.

Third, many conservationists believe dogs are a genetic threat to the wild types. Dogs can and do breed with wild wolves, including Ethiopian wolves, coyotes, all types of jackals, and dingoes. How do we know that such interfertility can take place? Mainly, one can observe it. Many a laboratory, including ours, has created and studied hybrids. With access to the North American and northern European wolves, people actually create and keep wolf-dog hybrids as pets.

The big question is: Does hybridization happen in the "wild" and, if it does, does it affect any of the wild types adversely? These are important questions and need to be constantly explored because if a wild genome becomes corrupted, the chances of uncorrupting it are very small. It would always be the smaller population that would be at greater risk of being swamped by a larger population. As we have said repeatedly, the dog population is 95 percent of the total worldwide *Canis* population. For a population such as the Ethiopian wolf, which retains only a few hundred animals, the risk of the wild species being overwhelmed by dog genes is close to inevitable.

How do we know when dog genes have escaped into the wild population and how many? Often the hybrid looks like a hybrid (plate 14). Hybrids commonly have characteristics of each parent. There is a good literature on crossbreeding of dogs with golden jackals, black-backed jackals, coyotes, gray wolves, and dingoes. They are also interfertile with one another.

Does this, however, happen outside the lab, in the wild? And if it does, does it make any difference? Throughout our studies, people have sent us photos (always of dogs) and asked if the animal pictured could be a hybrid. The dogs always look like they might possibly be a hybrid, but not exactly. A photo Alessia Ortolani sent us from Ethiopia suggests that some village dogs could be a cross with an Ethiopian wolf. Still, the possibility always exists that the wolfy-looking animal is pure dog.

Finding hybrid-looking animals in the wild population is rare in our experience. Watching wolves, coyotes, and jackals run around in the wilderness and seeing an adult wild type that looks like a hybrid pretty much never happens. As biologists, however, we have learned never to say "never." Right now world-class experts like Luigi Boitani, David Macdonald, and David Mech report hybrid dog-wolves living in the wild. Living and packing and hunting like wolves.

It is easy for a wild-type male to breed, at the edge of civilization, with a female dog. It is easy, if it is a season when the wild-type male has active sperm. The males of jackals, coyotes, and the different species of wolves are seasonal, having active testes for only a few months a year. Dog females can come into estrous in any month during a year. Thus, the frequency of male wolves breeding female dogs is only a fraction of what is possible by the ever sexually ready promiscuous dog males.

The wild types tend to be territorial when they are sexually active. Male coyotes, jackals, and the rest defend their territory against other males. Females defend the same territory from other females. It is the basis of pair bonding. The old truism that some of these animals mate for life simply means that the pair occupy the same territory during mating season for several seasons in a row. One must not forget that "for life" might be for three years.

Thus, the argument is that hybridization will most likely be the male of the wild type mating with the female dog. The female dog is promiscuous and could attract males from adjacent areas. Therefore, it is most likely that hybrid pups will be born in the dog's niche in and around human settlements. Those hybrid pups have an even poorer chance of surviving than the (pure) dog pups of the villages. Remember that the village dog pups have about a 4 percent chance of surviving to become reproductive adults. Female dogs abandon their pups after the nursing period and the pups have to find their own food. The competition among pups, and among pups and adults, is enormous.

The dog pups that do make it are either those pups born into a niche that isn't very saturated or those cute little tykes that can solicit food from a human. However, in our experience, hybrid pups tend to be spooky and shy of humans, even when raised in captivity with people and dogs. They show more fear responses and are more timid. Thus, what are the chances of the hybrid pup surviving to adulthood? Probably about the same as the stray dog — fairly low.

We have seen examples of female wild types bred by a dog, presumably because they could not find a territory and a male that went with it. The big problem here is she is going to have her pups in the wild and does not have a male to help her raise them. The pups are going to be shy of people. The reason for that is the wild-type pups form their social bonding patterns earlier than dogs — before the end of the nursing period. The mother wild type with no male to help will have human-shy pups that don't have much of a chance of survival.

Is it possible for a hybrid pup to be born in the dog niche and survive to reproductive age? Anything is possible; it is just unlikely. Is it possible for a hybrid to be born in the wild and survive to reproductive age? Again, anything is possible, but this scenario is also unlikely.

So why are conservationists so worried about something that hardly ever happens? How do they know that the wild types are being "corrupted" by dog genes? If you can't identify the hybrid on the ground, then there must be some way to test. Therein lies another problem. Geneticists tell us that hybridization happens frequently. One finds reports that 15 percent of the Ethiopian wolves or coyotes are carrying dog genes. Scientists are also worried that gray wolves are carrying a high number of dog genes. Some geneticists are saying they really can't distinguish between the species. Why? Because there has been so much admixture of genes (hybridization) that all the species begin to blend — which is just another way of saying that geneticists cannot tell the differences between the different species by simply looking at their genes. If that's the case, then early reports of Ethiopian wolves carrying many dog genes might simply be based on unproven identification techniques. No doubt the geneticists will come up with a methodology that will be able to distinguish between these closely related species, but some of us are not convinced they have done it yet.

Those conservationists working with endangered species have tended to believe the findings of these hybridization studies. Hybridization is worrisome with small populations, such as the Ethiopian wolves. Field biologists who are looking for a method of increasing population numbers of threatened species would like to maximize reproduction within the species.

Conservationists are, for the most part, trying to save the phenotype (the way the animal looks) of their endangered species. From the individual endangered animal's point of view, the effort is in getting its genes into the next generation. Thus for an Ethiopian wolf, what might be the best way of leaving your genes to the next generation might just be getting your genes into the neighboring dog population.

Still, we have colleagues who study populations of animals like the Ethiopian wolves and Italian wolves and think wild animals are corrupted by dogs. It may well be that in areas like Italy, where the wolf niche is not saturated, or in areas where the wild niche is so changed by human activities, that a hybrid might survive. Great biologists like J. B. S. Haldane, disbelieving Darwin's idea that change came about gradually—through natural selection and survival of the fittest—thought there were quicker ways for evolution to take place, such as through hybridization. Many biologists believe that hybridization is an accelerated way to get started in adapting to a new niche, since it rapidly produces genetic and phenotypic variation that can then be selected for or against. Certainly, most modern breeds of dogs were created by crossbreeding (hybridizing) two different breeds together. Crossbreeding creates new and interesting phenotypic and behavioral shapes quickly, which is how we got those hundreds of breeds of dogs in just a few hundred years.

There are, in short, a number of questions that scientists need to ask themselves for which they do not, as yet, have definitive answers. Such as, is it, in fact, possible for dog genes to stray into the wild? Is it possible that such a leakage will have a deleterious effect? Or is it possible that the leakage of dog genes into the wild could help an endangered species survive? In sum, is hybridization actually happening (we are not quite sure) and, if it is, is it necessarily bad?

PART IV

Summary

16

Where—and Why—Are All These Dogs?

The hypothetical representations of how the populations of wild types, and then the emerging village dogs, were distributed in the world at four different times in their histories are depicted in four diagrams here. The first illustration (fig. 16, *top*) is the world of wolves, coyotes, and jackals before there are any dogs. The diagram represents the distribution of these animals only and not the numbers, but the educated guess would be that the smaller animals would have a higher density per unit area. In other words, even though coyotes occupy a smaller area, there could be more of them than there are wolves.

Of course, wolves, coyotes, and jackals aren't a uniform size and shape over their entire range or over time. The Arabian wolf is about a forty-pound animal, while the arctic gray wolf can be over a hundred pounds. Theoretically, animals can change size in situ over long periods of time.

Sometimes scientists name variations in a species and call the local population red wolves, or Mexican wolves, or

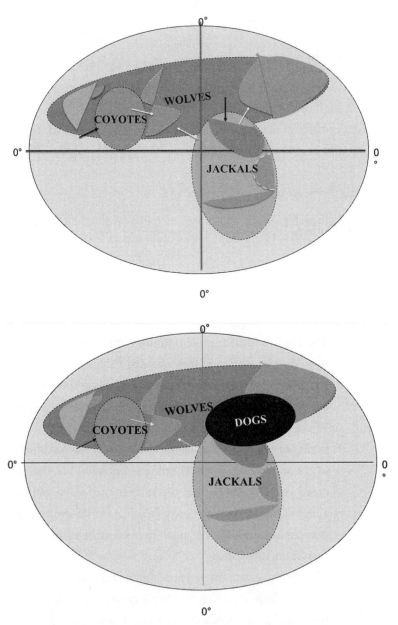

Figure 16. *Top*, Distribution of wild canids before the appearance of dogs. *Bottom*, Dogs appear as a population some 7,000 years ago.

Arabian wolves. These variations are a subspecies. In other words, the alleles that make a wolf red, for example, tend to be clustered at a higher frequency in the southeastern portion of North America. Thus, in the diagram we show the patchwork of locally different subspecies across the range of wolves, coyotes, and jackals.

For whatever reason, red wolves could be better adapted and produce more surviving offspring than gray wolves in the southern part of the United States. Or perhaps a founder effect exists there, where the population of wolves in that southeastern region was wiped out by a dreaded disease and just by chance alone only one pregnant red-colored wolf remained. She and her seed could have repopulated the whole region.

At the edges of these various areas containing subspecies, the animals can breed together and produce offspring. The offspring of such matings are sometimes known as hybrids. The zone of hybridization is often called a suture zone. Genes can "flow" throughout all the different populations—one to another and to another. Over time, genes flow from one population to the next. In populations that border one another, the flow is faster than it is within widely separated populations. In other words, genes can and do flow faster between red wolves and gray wolves—whose habitats are on the same continent—than they flow between gray wolves and Arabian wolves.

Another caution attends our diagram. Wolves and coyotes and jackals all acquired their scientific names (*Canis lupus, Canis latrans, Canis mesomelas*, etc.) in the nineteenth century. In that period, many people thought the world and all the animals thereon were created by God in 4004 B.C. That was before Darwin realized the earth was much older than anyone had imagined and that species such as wolves, coyotes, and jackals were ever-changing adaptations to a niche. The niches were constantly changing, and the species were constantly readapting to those changing niches (plate 16).

To illustrate that point, we put the wolf population farther south in figure 16 (*top* and *bottom*) than in figure 17 to indicate that before dogs, beginning some 20,000 years ago a massive glacier covered the top of the earth and that the glacier extended farther south in North America than it did in Eurasia. A brave explorer could have walked from New York City (40° north) to Dublin Ireland (50+° north) on glacial ice. That is why we tipped the wolf population farther south and tipped the diagram that way.

After Darwin, most biologists thought a species was a population of animals that is sexually isolated from all other species, and since wolves, jackals, and coyotes are not sexually isolated, they are not different species. Most often, well-defined species have a different number of chromosomes or a different arrangement of chromosomes. Humans, chimps, and gorillas have a slightly different number of chromosomes, and there are some gene sequences that have been inverted. The fertilized egg usually cannot deal with these differences, and the offspring dies at some time during embryogenesis. The horse and the donkey, in contrast, are great examples of closely related species with different numbers of chromosomes that can mate and produce offspring — mules — but the offspring are sterile, thus reproductively dead.

Wolves, jackals, and coyotes have the same karyotype (chromosome count) and can mate easily, and do, wherever populations overlap (such as in a zoo or in the New England wild). For a paleontologist, they are technically the same species, just as a Tanzanian Maasai and a Canadian Inuit are the same species, even though they look different from one another.

Remember Joseph Chang's argument, cited in chapter 1, that fifty-four generations ago, any sexually dimorphic creatures on earth had a quadrillion great-great-grandparents. All the humans on the planet had the same great-grandparents some fifty-four generations ago (which might be as little as 1,000 years ago). Applying that argument to wolves, coyotes, and jackals, they, too, all had to share the same grandparents fifty-four generations ago — which might be as little as 200-odd years ago.

The problem with Professor Chang's argument is the assumption that the individuals in the population breed at random—that is, anybody has an equal chance of breeding with anybody else in the population. But it is hard to see how that could be. In our discussion of the bowl of marbles (chap. 4), we said that any marble could mate with any other marble in the bowl but that it was difficult to move around the bowl and any one marble was most likely to mate with the nearest marble.

If, as we suggested, all the marbles in the bowl died except one pregnant marble, then in all succeeding generations she is the common ancestor. In that case, one doesn't have to compute fifty-four generations. Chang's hypothesis suggests that all individuals leave a surviving lineage and, of course, they do not.

The wild types of the genus *Canis* are not sexually isolated and, technically, if you're a paleontologist, they are the same species. Well, not exactly: though they may be interfertile, they are adapted to different niches. That is, they live in different "bowls," and, therefore, to an ecologist they would be different species—as long as they stayed in their own bowl.

They are local adaptions to their environment. Indeed, in terms of skull measurements, coyotes and jackals are more alike than either of them are like wolves. If you took a big jackal skull and buried it in New Mexico, the experts would have a hard time simply classifying it as another coyote. That might lead to some thought-provoking questions. If jackals were taken from Africa and released in New Mexico, would they survive? Would they breed with coyotes? Would they become an invasive species and replace coyotes?

Thus in the second diagram (fig. 16, *bottom*), we suggest a gene flow between the so-called species of wolves, coyotes, and jackals. In the color section of plate 16 is Lee Spector's conception of a Darwinian tree of the genus *Canis* with genes flowing through time, evolving and reevolving into one form after another in the same place. Whenever a geneticist looks at any of these species, they always report a number of dog genes, which is how the concern over the Ethiopian wolves got started—with the claim that they were carrying dog genes. The red wolf, the coyote, the gray wolf, and the New England canid are all also reported to have dog genes.

Can species change size and shape in the same place—the same geography—over time? When the last glacier covered the top of the world, all climates were different from what they are now. In North America 20,000 years ago, the tundra went all the way to Texas and Georgia and, in Europe, to France and Spain. Is it surprising that one can find skeletons of wooly mammoths over that entire area now? Is it surprising that when the tundra went away, the wooly mammoth did too? Is it surprising that when North America was mostly tundra you had really big wolves that we call dire wolves? Is it surprising that we didn't have any coyotes, which show up after the glacier retreats? Some authors, who believe in the fixity of species, believe that the coyote must have come across the land bridge between Asia and Alaska. How else could they have traveled here? Certainly they couldn't be an adaptation of some wolflike animal that was already here adapting to the changing climate—or could they?

Behavioral ecologists who study Ethiopian wolves think of them as a species that occupies that geographical niche. Indeed, they not only think of them as a species, they are worried that that shape (the species) is going extinct and they would like to stop that from happening. They work with local people to pass new laws that protect the wolves from being shot and trapped. They work to preserve the wolf habitat—their niche. They also try to reduce the wolves' contact with other members of the genus so that diseases that might spread among them and hybridization that might corrupt their gene pool with nonadaptive genes are inhibited.

Inhibiting the spread of disease and the incidence of hybridization is a fancy way of saying that the behavioral ecologists try to keep the wolves from breeding with dogs. Conservation programs support sterilization of local dogs to keep them from breeding with the Ethiopian wolf. Similarly, the people who reintroduced the red wolf into South

Carolina are trying to save the "species" of red wolf from extinction. The first thing they had to do was remove all the coyotes and dogs from the reintroduction area so they wouldn't corrupt the newly introduced animal's gene pool. For a while they had a plan to test each new generation of red wolves to make sure that no coyote or dog genes had snuck in.

Thus, what many conservationists do is try to protect the form or shape from changing. They are, in a sense, trying to inhibit evolution.

Enter the Dogs

Our next diagram (fig. 16, *bottom*) has a new population, meaning a new shape, evolving. In the Middle East 7,000 years ago, a new population appeared, depicted on cave walls and pottery—what we know as dogs. Did dogs start their evolution in the Middle East? No evidence so far is conclusive. Were there dogs before that? Well, probably. The archaeological evidence shows—here a skull and there a skull, but not necessarily a dog skull. The investigations contribute fun arguments and discussions. But so far paleontologists, archeologists, and geneticists do not have a good methodology to tell the difference between dogs and wild types. Why did they suspect in the first place that dogs descended from wolves, and if they did, in fact, descend from something that we now call a wolf, which subspecies of wolves were they, and where were they in the first place?

What we do know is that, about 7,000 years ago, drawings first began to appear on pottery and cave walls of animals interacting with people in doglike ways. Some of the animals have curly tails and collars or were depicted in a house or chasing another animal alongside people.

For our story, a first dog existing before a *population* of dogs doesn't really matter. For an evolutionary biologist, there possibly is no such thing as the first dog. Adam and Eve dogs are the way creation occurs in religious texts but not in evolution. Evolution is the change in shape of individuals within a population as an adaptation to a niche. That changed population appears 7,000 years ago (fig. 16, *bottom*).

In the years that followed, some variations developed within the regional populations. Selective pressure in any location will produce some nonrandom distribution of alleles. A genetic nonrandom distribution of dog alleles can occur in three major ways:

1. *Adaptation to local conditions by natural selection.* An illustration of how that works is seen in the relationship of body size and shape to latitude and altitude. Dogs vary in size, shape, and coat texture over their entire range. Like their wild cousins, they tend to be smaller, or basenji-like, in equatorial and hot regions, grading to larger husky-like animals as latitudes and altitudes increase. Such size variation is common among a variety of species and, as we have pointed out, the variation is generally known in science as Bergmann's rule. The importance of Bergmann's rule for dogs is that it supports other evidence that they are in large part a product of natural selection.

2. *Founder effects produced by fluctuating population numbers.* A disrupted but recovering population will not have the same frequencies of alleles (genes) as the previous population. Nonadaptive differences in regional phenotypes can be the result of founder effects. Characteristics such as ear carriage or curly tails, or a high frequency of a particular coat color can also be the result of a founder effect.

3. *Postzygotic culling by humans.* Dogs have an enormous number of pups all year long close to villages. In many areas, people destroy pups, in part, for reasons of population and pest control. Regionally, humans have preferences for certain characteristics and will spare pups with the preferred color, ear carriage, tail carriage, or whatever. The culling is capricious, selecting for superficial characteristics described as neutral to natural selection—selected neither for nor against.

One of the big points about postzygotic selection is that in many cases the pup's survival depends on becoming adopted by a human. Humans are quite often attracted to novelty. Alan Beck pointed out years ago that people differentially fed the dogs of Baltimore, often favoring one or two coat colors over others. In doing this, they were inadvertently selecting for some other characteristic that they were not aware of. We saw this happen in white livestock dogs, discussed in chapter 13.

The fact that there were dogs several thousand years ago and that they varied means neither that humans domesticated the dog nor that humans were selecting for the variations by breeding selected individuals. There is no evidence of either at the present time. It wasn't until roughly Roman times that there is a hint that breeding for shape or behavior might possibly have started. Even then, the wording of ancient manuscripts isn't clear. The ancient Roman writer Marcus Terentius Varro says if you want a hunting dog, get it from a hunter and if you want a sheepdog get it from a shepherd. That implies there was some kind of selection going on, but it isn't explicit. It could be that the early behavioral development of puppies was important to their adult duties, and thus pups growing up with shepherds and their sheep were more likely to be good sheepdogs than dogs that grew up in hunting communities. There is also the possibility that shepherds and hunters were disposing of pups that didn't work well—postzygotic selection.

We assume that the new dog shapes were adaptions to variations in the geography or the niche. We also assume from that niche-adapted shape there was a leakage of genes between the various dog communities and a leakage back into the wild critters.

Why not? It happens all the time now where dogs breed occasionally in the suture zones with jackals, coyotes, and wolves—in the wild—and wolves breed with coyotes. A large number of Australians think the dingo is just a dog. They are wrong, but dingoes do breed with dogs. If dogs now breed with all the wild types—in the wild—why wouldn't they have been breeding with them for the last 7,000 years?

By 2,000 years ago (fig. 17, *top*), dogs were every place humans were, which covers most of the world. It may be that dogs hadn't progressed south of the Sahara at that time. The Sahara seems to have stopped the expansion of dogs into southern parts of the African continent. Some writings or wall drawings suggest that dogs finally reached places such as Namibia and South Africa with the arrival of the migrating Bantus 1,500 years ago.

Prior to 2,000 years ago, no solid evidence has been found that humans were affecting the breeding of dogs. Nothing in Homer or the Bible suggests kennels and selection for breeds. That doesn't mean that a pharaoh didn't have a kennel full of something that might look like greyhounds—there just isn't any evidence. And even if a dog-sport-loving king somewhere was breeding his dogs, it wouldn't have had any effect on that great big worldwide population of unrestricted dogs.

Why doesn't the gene flow between this massive and far-ranging dog population and the wild types destroy the wild types? To put it another way, how can a gene flow exist between dogs and the wild forms without causing them to evolve into some common form? The worry now is that if dogs breed with the red wolf or the Ethiopian wolf or even the gray wolf, they will corrupt those animals, causing the current forms to become technically extinct, as described in chapter 15 and elsewhere. The wild forms are protected from this, however, by the existence of the niche. There is still a wolf niche, a coyote niche, a jackal niche, and a dog niche, and the hybrids between any of them will not reproductively survive as well as the niche-adapted forms. In short, the hybrid will be selected against.

Evolution of canids can be thought of in terms of the Darwinian evolutionary tree. At some time in the past, coyotes and wolves diverged, splitting into two genetic lineages. Then, at a much later time, the dogs split off from wolves. But as the song says, "It ain't necessarily so." Lee Spector's tree (plate 16) is more like what most of us believe happens. Over time, the genus *Canis* changes constantly and locally with new shapes and sizes emerging and going extinct and reemerging

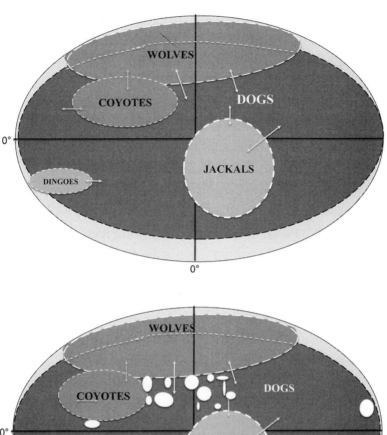

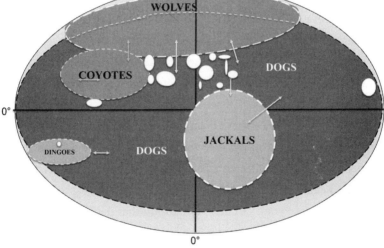

Figure 17. *Top,* From 7,000 years ago until 2,000 years ago, the dog population increased in both distribution and population numbers. *Bottom,* From 2,000 years ago until the present, there appear these sexually isolated populations of dogs called breeds. The vast majority of these breeds have appeared in the last 200 years.

and grading into one another as local conditions change. Niches are not constant either: they change continuously. What we think of as being a species now was different in the past and will be different in the future. Evolution is the process of species continuously evolving to new and changing niches.

Oh dear! How far does one have to push this? If Inuit people are adapted to the arctic tundra niche and the Maasai are adapted to an African semidesert scrubland — then are they different species? Maybe to an ecologist they are. That seems a little politically incorrect to even talk about at the moment. Thus the first question would be, when discussing evolution, which species or collection of species are you talking about?

It is also worth noting that the word "hybrid" is used in many different ways. Entities such as hybrid sheep or hybrid corn are crosses between two varieties of the same species. A variety can be a breed or it can be a subspecies. A coyote-dog hybrid is thought to be a cross-breeding of two species. A wolf-dog hybrid was thought to be a cross between two species, but taxonomists have changed the species nomenclature for the dog to *Canis lupus familiaris*, which indicates they are, at least currently, just varieties of the same species. If you are an ecologist and believe a species is a niche-adapted form, then in spite of whatever name taxonomists give to the dog, the ecologist still thinks of wolves and dogs as different species because they are adapted to different niches. The problem with the different species argument is that in the back of everyone's mind is the notion that people domesticated dogs from wolves and they are now a human responsibility.

What difference does it make if the wild types are classified as different species? Our friends who are trying to protect a species like the Ethiopian wolf care a lot. If, however, you were an Ethiopian wolf and your job was to get your genes into the next generation and beyond, then it just might be that, right now, you would have a better chance of passing genes on if you mated with dogs. Blasphemy!

As mentioned earlier, the biologist J. B. S. Haldane suggested that one of the major ways of creating new species is by hybridization. Maybe the New England creature referenced at the very start of this book—a hybrid wolf-coyote-dog—is on its way to becoming a new species, adapting to a new niche in New England.

Whatever problems hybridization between wildlife and dogs might cause, the impact of hybridization on wildlife is still probably not as significant as the effect of dog diseases. Dogs constitute such a massive population compared to the wild types that an outbreak of any disease in dogs anywhere in the world affects wildlife. All the specific wild types added together equal about only 5 percent of the total dog population. The total of all the wild types added together consists of about 50 million animals. Certainly, the wild types are such a small percentage of the genus *Canis*, that, in numbers, they are engulfed by the billion dogs (fig. 17, *top*).

In the past 200 or more years, people have regionally captured dogs and isolated them sexually from other dogs. Often, after capturing dogs, people took them far from where they had evolved. The little white islands in our diagram (fig. 17, *bottom*) are breeds of dogs. No matter where dogs were captured, they tended to be isolated in homes in North America or Europe or other "developed" nations.

For the most part, breeds of dogs are insignificant in the total dog story. They are artificially isolated, mostly by the world's richest people. They tend to be selected for non-wild-type conformations. They are bigger or smaller than the adaptive form. Further adding to the stress of the purebred dog, many are bred for a deformity such as short legs, pushed-in faces, or extra-long bodies.

Some of these dogs stray, and their genes do escape back into village dog populations. But there is little evidence that this has much lasting effect on the larger population. As with any hybrid offspring, they tend not to have the adaptive shape and are probably less likely to survive. Thus the effect of breeds of dogs on the world population of dogs is probably trivial.

17

What Should We Do—
If Anything—with All the Dogs?

We will try to summarize a case for the evolution of dogs *as dogs* in another way. Based on archaeological evidence, we assume that many thousands of years ago a new niche appeared, created by people as they altered their environments in the process of moving from hunting and gathering to agricultural societies. With agriculture came a great number of permanent settlements called villages, many of which became large villages, which we now call cities.

That is not to say that before the age of agriculture various permanent settlements did not exist. People spent generations on seashores beside huge shellfish fisheries. Great mounds of shells accumulated in places like Florida. In areas where game animals were plentiful, people could stay in one place somewhat permanently. In Namibia, with a class of Hampshire College students, we explored a water hole in a rock outcropping that was the only drinking water within miles of a gigantic grassland. In ancient times, hunters spent time on the rocks and etched pictures while waiting for their favorite dinner to come by for a drink. Archeological finds suggest long-term occupancy in some spots that animals frequented and where they could be trapped in some easy way. There are so many easier ways to catch game than chasing them with a dog.

Could some wolves or jackals have evolved into dogs in these pre-Neolithic or even Mesolithic villages? It is possible of course. But so far there is no evidence of a population of dogs until well into the Neolithic. Our guess is that 7,000–8,000 years ago is the earliest date that verifiable evidence confirms a population of dogs. The bet is they weighed thirty pounds and looked like what are today called mutts.

In the age of agriculture, the evidence of village life becomes much more frequent. In addition, this new agricultural economy was locally messy. Ancient villages are wonderful places for archeologists because the inhabitants stayed in the same place for generations with all their junk accumulating around them. The detritus is often many feet thick. As the human population grew and people became more dependent on grain crops, they tended to cook them in one place. The waste in that place increased in proportion. The crops and the garbage produced from crops attracted many, many species of mammals, birds, insects, and other fauna.

Fredrick Zeuner wrote, in his book *A History of the Domesticated Animals*, that all our domestic animals with a few exceptions were originally crop pests. We would suggest that his title should say "domestic" and not "domesticated." He should have pointed out that the domestic animals he did not deem crop pests were probably trash pests. These non–crop pests were species such as rats, mice, chickens, turkeys, and the dog. The exception to the pest/scavenger role may be the cat, which hunted domestic rats and mice in the domestic pest-ridden environment. The more mess, the more "domestic" animals.

Early on, people saw these various species of crop pests and village scavengers as a great source of food, and some people still do. Just as their ancestors waited at the water hole for game to come to them, the agriculturalists had the grain crop and associated waste thereof to attract edible animals to them. The wild turkeys in our farmyard eating corn are pests, but come Thanksgiving we might look on them as a major part of a meal. As modern hunters, we have the turkey eating out of our hand and after a bit it isn't hard to grab one (plate 17).

E. O. Price pointed out years ago that whatever genetic changes defined our domestic animals they still have to be tamed in every generation. With dogs we refer to this taming as the critical period of socialization. But those that raise cattle, sheep, goats, chickens, or whatever know that each generation has to be socialized with people or they grow up wild. Range cows from Texas scrubland are wild and tough to manage, while the same species in India pull plows and carts, and their calves graze around houses, have names, and will come when called. Almost any species can be trained to do atypical tasks. Ostriches have been trained to pull carts, and some professionals teaching people how to train dogs start by teaching them to train chickens.

Domestic animals were and are attracted to human resources. Domestic animals are species that can eat in the presence of people. They are not genetically tame but are easily tamable for whatever reason. The dog isn't all that different.

Deer, moose, squirrels, rabbits, and woodchucks all consider our yard a place to graze the fields, the lawn, and our apple and peach trees and are relatively easy to catch. Perhaps Neolithic farmer/hunters have done that all along. Indeed, peoples around the world have found that the best hunting sport with dogs is hunting pests. Fox hunting, coon hunting, rats, rabbits, genets, and green monkeys are all on our hunted list. The only reason to hunt with dogs is for the sport of trying to catch something with a dog. Like most sports, it is hard work and takes a lot of time. It certainly is not a great way to conserve calories, but it is a fun way to control pests if you can afford it.

Pure breeds of dogs are represented in figure 17 (*bottom*) by the white islands. Breeds are referred to as products of artificial selection. Within the huge population of dogs, some are captured by people and sexually isolated from the rest of the dog population. Each white island was created by eugenics, which is a nonrandom breeding for a phenotypic characteristic that complies with some morphological or behavioral standard. Dogs caught and isolated and bred for a capricious character such as coat color are most likely genetically doomed. Unless their genes can escape back into the larger population, they will eventually succumb to inbreeding depression.

Arrows in the diagrams point to the leakage of genes from one population to another. Note that the leakage of genes between the various dog populations also flows back and forth between the dogs and the wild types.

Are these abstract diagrams in figures 16 and 17 a representation of how dogs evolved? Probably. If we traveled around villages in Greece in Homer's time, would the diagram have been representative of the dog world then? Not exactly, of course. There weren't as many people as there are today, and thus there couldn't have been as many dogs as there are now, either. In Homer's books, however, there were a lot more people lying around dead from their constant wars — and dogs eating them. It is doubtful those dogs feasting on the battlefields were purebreds.

For a behavioral ecologist, the diagrams are a wonderful way to illustrate an ecologically and evolutionarily heathy, robust species. Any genetic disease will be quickly selected against — except in the little white islands.

This is not to suggest that people should not have their little white islands of dogs. But ethics would suggest that people shouldn't select for those capricious traits, which are deformities that put the animal in constant pain or some daily stress such as difficulty breathing.

Why should we do anything about all those village dogs? If the village dog is, as we suspect, similar to the several other wild species that scavenge from human beings for their diet, why should society be persecuting them for filling the niche into which they evolved?

Of course, any wild species that lives in large numbers around humans must be monitored or controlled in some way that protects human health and safety. Dogs affect the quality of human life in many ways—they bite, they bark, they spread lethal diseases, and they are nuisances in uncountable ways. In many places in the world, dogs are routinely poisoned or shot. When dogs appear to be sick, showing signs of disease, local officials create eradication programs.

Other species besides dogs have evolved or adapted to feeding in the human domain. The list is wonderfully long and includes not only rats, pigeons, cockroaches, and bedbugs but also vultures, crows, raccoons, and, now, coyotes, jackals, and coyote-wolf crosses. All those melodious birds that visit our feeders and nest in and around our house and barn are more than welcome, although we suspect the phoebes that nest on our porch bring lice. When we were actively farming and kept livestock, we became home for any number of cats that volunteered for service in our barns. At one level, they are all fascinating, and at another, they are all pests. Ever since we lost our last dog, we are routinely visited by wildlife. Not only the turkeys and deer on the lawn and the field but also bears plunder and damage our bird feeders. (A colleague says bear tastes awful.) It is hard to fess up and admit, but those lovable dogs we have had all these years have put serious pressure on local wildlife. In many ways, life here on the farm is more interesting without dogs. For example, with no dog to warn off, we don't mind all those porcupines in the yard, which are curious in their own right. (One neighbor says they taste good.)

Except for the mice and moles, all the wildlife in our yard make somewhat good "pets." We have entertained coyotes and raccoons, and we know a few people who have kept foxes. Last March, two big old vultures showed up at the empty feed tray after they had wintered farther south. They must be old friends from last year, or how else would they know that the tray is a vulture feeder? It was empty, but they came and checked it out. We feed them road-killed animals or birds, and our grandson has trained the school bus driver to stop so he can pick up carcasses in the plastic bags he carries in his pocket. The grandson also has a field camera set up at the feeder so we can keep track of what is going on. When the fishing is good, the vultures and crows eat the parts of the fish we don't.

The prevailing claim all around us, however, is that dogs are different from the other critters that frequent our yards. They are, after all, allegedly man's best friend. What a hollow statement that can be. It often seems just a way to sell dog food or keep veterinarians working or raise contributions for humane societies. In the current popular mythology is the perception that dogs were domesticated by humans to do something useful. Thus, stories full of euphoria abound about working dogs that help us. Almost all those stories are about people in first world cultures that can for example afford $30,000 dollars to train a guide dog. The guide dog and blind person connection is not exactly mutualism because it doesn't have much if any benefit for the sterilized and constantly controlled dog.

It is assumed that dogs and people live together for each species' mutual benefit. It doesn't matter to the conventional wisdom that there is no evidence for that.

The argument for mutualism is a strange one. It goes something like this. Humans "domesticated" the dog to be a companion for hunting. Therefore, if any dogs are not pets and/or companions, meaning they are not totally dependent and restricted, then they are strays. Strays, it is commonly thought, are the result of irresponsible behavior of people who were their companions. Thus a humane society should be placed out there and accept contributions to employ people to promote a "humane" treatment to rectify the problem.

An outgrowth of this approach is evident from the numerous programs worldwide that sterilize village dogs. Since most village dogs live in warm countries, the best time to go there and sterilize them is during the northern winter months. Many of these programs are located in tropical paradises such as the Bahamas and the Caribbean islands, where the programs are carried out in January. Spain, Italy, Portugal, and other Mediterranean countries are popular wintertime destinations for northern Europeans for the purpose of capturing and neutering and, even worse, removing the dogs from their niche and sending them back to developed countries for adoption.

These programs are designed to reduce the population of dogs without removing individual dogs. Many of the dogs are captured, spayed or neutered, and released back to their street or village. In human medicine if a heathy organ is removed with no benefit to the individual it is termed mutilation. However, the term tends not to be applied in areas where domestic animals are commonly spayed and neutered. The philosophical question could be asked whether the village falls under the same set of values as animals being raised for food or pet animals where sterilization is a convenience for the owner.

The idea that one could reduce populations through sterilization of adults was originally suggested by Edward Knipling in 1938, and it was still an exciting idea when we were in school in the 1950s. There were rules, however. In Knipling's technique with blowflies (an agricultural pest), the female insect mated with one male only. That male had to be sexually active even though sterilized. They were sterilized by irradiating them, a procedure that was touted at the time as a peaceful use of atomic energy. Other restrictions stated how and where the method could be employed. The area where the sterile males were released had to be reasonably well isolated from other populations of the target species to prevent migrations of fresh populations into the treated areas. The sterilization system worked well on blowflies on peninsulas like Florida but not in open spaces such as Texas.

The question now is, will sterilization work on reducing village and street dog populations? Unlike the blowflies, both male and female dogs have to be sterilized because both are promiscuous. The blowfly females are monogamous. And unlike sterilized dogs, blowfly males are sexually active.

For dogs, there might be an immediate benefit from sterilizing large numbers of them locally. By sterilizing large groups of adult dogs and putting them back into the population, they will still be competitive with growing juveniles. That means the death rate of juveniles should be about the same after sterilization of adults as before sterilization and remain high until the adults begin to die off. The sterile adults will still form a barrier preventing fertile dogs from moving in from the periphery. Furthermore, the sterile adults won't produce a replacement population. The problem with the technique for dogs is that in a normal population only about 4 percent of newborn pups need to survive to keep the population stable.

The estimate is that adult street and village dogs have a four-year average life expectancy. The population of sterile adult dogs, given a normal age structure, will begin to drop almost immediately and will decrease by better than 25 percent per year. The drop in population will leave an opening in the niche for replacements. Replacement animals from adjacent areas can move in, and/or juveniles will have a better life expectancy. Luigi Boitani and his colleagues studied the feral dogs at Rocca di Mezzo in Italy, where no successful reproduction was apparent and yet the population remained stable for years, with emigration from the periphery. Our guess is that sterilization won't create much difference in the population unless it is massive and continuous—meaning the next 100 or 200 years—which is fine as long as the society has the kind of excess money and time to keep at it.

Sterilization is thought to be more humane than killing. But from a population point of view, neither killing nor sterilizing will make much difference in the long haul. That might be a bit of an exaggeration because if you reduce the population of dogs for some period, then the niche becomes available for another species. The illustration we used earlier (chap. 7) was that the demise of vultures in India provided an opportunity for dogs to expand their range.

Some cities, such as Rome and Istanbul, are full of cats, which have replaced the dogs after a dog control program. In most northern American cities, the dog problem appears to be under control, and some unusual new animals are lurking around at night. The recent discovery of a coyote-wolf hybrid operating in cities from Detroit to New York is attracting attention. Soon we will need a control program for them, just as we need control programs for other wildlife species.

Some humane programs capture village dogs, neuter them, and send them off to shelters in rich countries to be adopted into family status, where they are made totally dependent and entirely restricted. They get to walk on leash for exercise, and they are walked occasionally to a doggie park. The so-called benefit to the dog (a social animal) is not measured in terms of a better social life for them but, rather, in terms of longer, healthier lives.

For many street or village dogs, the problem with being inserted into this life-altering experience is that they lack the ability to adapt to the new environment. Behavioral problems accompany that lack of adaptability. Why? Well, they weren't socialized from early puppyhood to that restricted new environment. By the time they get to and through the shelter and into a home as a companion animal, their critical period for social development is long past.

It is common for working dogs—military dogs, customs dogs, police dogs—to be retired and confined to a domicile—where they sometimes can display abnormal and unacceptable behaviors. A major problem known as separation anxiety appears in many of them. The military and police often just euthanize their dogs rather than passing them on to a new, totally restricted environment. The same situation is true with dogs kidnapped from their little island paradise somewhere and forced into the restricted dog category. Often we hear professional dog trainers complain about these dogs. To make the transfer work at some level, the new owner must be taught how to manage such a dog. It's a complicated, often long-term, mission.

In these adoption programs, little consideration is given to the reduction in the quality of the lives in which the dogs have lived comfortably. Like wild animals, village dogs in their natural environment are fine-looking individuals with rich social lives. The males are sexually active all year long. The females are promiscuous, and some have robust litters every ten months. These remarkable dogs are captured, mutilated, and taken off to lead an ostensibly dull life as a house pet or companion dog, commonly devoid of contact with other dogs. They are reduced to being totally dependent and totally restricted in their movements. In some ways, they seem like the poor people in nineteenth-century paupers' prisons in England who weren't allowed to breed because they were regarded as genetically inferior simply because they were poor.

Some of us think such dog-limiting programs are cruel. Some of us think these programs don't affect the local dog population numbers very much, and even if they did, the programs would have to be carried out forever simply because the niche created by humans is still there, and the dogs are the best species to occupy it.

Third world governments often are eager to attract help in reducing dog populations. For governments, dogs are pests. But it is rare in such governments to find someone who understands the population dynamics of pest control. Sending their local dogs off to first world countries has the feeling of population reduction and the feeling of lessening the danger of disease, biting, and nuisance problems for people, and a halcyon future for the dogs. But is this short-sighted procedure really better for the dogs?

So, what to do with all the dogs? As we have tried to show in this book, the village dog is a wondrous animal with a self-tailored life style that suits it just about perfectly. It is well adapted to its niche. If people require fewer dogs in the environment, perhaps the most humane way of reducing populations of village dogs is to shrink the niche to which they are adapted. A smaller niche would support fewer dogs.

ACKNOWLEDGMENTS

William Dietel, Helenty Homans, Charles Longsworth, Susan Mellon, Adele Simmons, the USDA, and Wolf Park provided a great deal of wherewithal for years, building large kennels and supplying a special classroom, office, summer salaries, and research space. Hampshire College students were able to join our studies and learned how to carry out field projects. They became eager researchers in important projects. Skip and Marilyn Harned and the Anatolian shepherd dog club sent us (including our dean at Hampshire, Susan Goldhor) to Turkey to acquire dogs for our livestock-guarding dog project and for their rare breed initiative. Several trips to East Africa were helped by Abigail Shearer, herself a Hampshire alumna.

Our kids Karyn and Tim fed and watered dogs (and sheep) for years and assisted in the experiments, learning along with the students about dog behavior. They were great traveling companions when we went looking for dogs in faraway places and took care of one or the other of us in the field on long expeditions to Argentina, Denmark, East Africa, Ireland, Italy, Portugal, South Africa, Turkey, Venezuela, and the former Yugoslavia. Karyn trained and raced a small team of sled dogs, and Tim trained and trialed his own bird dogs, which gave us a different look at dog behavior.

Our colleagues at Hampshire actively participated in the studies, making up for our own shortcomings. Mark Feinstein coached us with learning and cognition, Lynn Miller taught us genetics, Lee Spector with Cristina Muro and Ramon Escabedo Martinez created the testable models and did the math, and Mike Sutherland designed the statistics for our specific experiments. All of these wonderful scholars coauthored papers with us. Mark is the coauthor of *How Dogs Work*, about the ethology of dogs, which the University of Chicago Press published in 2015 and is a companion book to this one.

Many of the thirty-five years of students who studied behavior as undergraduates at Hampshire stuck with it through graduate school, some becoming coworkers in the scientific world — Cindi Arons, Abby Drake, Gail Langeloh, Jay Lorenz, Kathryn Lord, Jeff Maguire, Alessia Ortolani, Jonathan Sands, Michael Sands, Dave Schimel, and Richard Schneider carried out milestone experiments that have changed the way we think about the behavioral ecology of dogs.

Throughout this book, colleagues such as Luigi Boitani show up over and over, which indicates how important he and they were to our thinking. Clive Wynne and his wonderful students, especially Monique Udell, carried out experiments comparing the cognitive abilities of dogs and wolves, which is so important to our thinking.

Many people contributed photographs from very out-of-the-way places. Some actually searched for photos of the perfect scenes and also the right number of pixels. Most helpful on this were Jane Brackman, Virginia Dare, Daniel Stewart, Alain Weiss, and Kristina van Haagen. We are hugely grateful to them all. Daniel does his own field studies and he, along with Matthew Berry and Mike Wood, have contributed to our knowledge of the wonderful African village dogs. Photographer

Monty Sloan of Wolf Park seems always to have the right image. Tom O'Dowd also of Wolf Park translated dozens of our old-fashioned analog images into digital ones at Color Images in Chicago. Dave Dumas of Pivot Media in Florence, Massachusetts, is the layout artist that puts them on the page. We received immediate assistance with graphs and tables from Kathryn Lord, former student and now colleague, and Dave Pinardi, former little kid playing with our son and now designer and owner of Adagio Digital Arts + Communication.

It was a miracle that we could put two books (*How Dogs Work*) and this one together in a year. It couldn't have been done without the constant support of our agent Malaga Baldi. The other miracle of support came from the University of Chicago Press: Carrie Adams, Christie Henry, Amy Krynak, and Yvonne Zipter who (1) put up with us and (2) were ever-present with advice and simple solutions to our many problems.

Studying dogs the way we have done has been a dream job. We met so many people who put us up and put up with us. We made tons of friends. All of you made this story possible, and we hope it all makes sense to you. Thanks again and again.

BIBLIOGRAPHY

Preface

Coppinger, L. 1977. *The World of Sled Dogs.* New York: Howell Book House.

Coppinger, R., and L. Coppinger. 2001. *Dogs: A Startling New Understanding of Canine Origin, Behavior, and Evolution.* New York: Scribner.

Coppinger, R., M. Sands, and E. Groves. 1973. "Meet New England's New Wolf." *Massachusetts Wildlife* 24, no. 3: 8–11.

Coppinger, R., L. Spector, and L. Miller. 2009. "What If Anything Is a Wolf?" In *The World of Wolves: New Perspectives on Ecology, Behaviour and Management,* ed. M. Musiani, L. Boitani, and P. C. Paquet, 41–67. Calgary: University of Calgary Press.

Francis, R. C. 2015. *Domesticated: Evolution in a Man-Made World.* New York: W. W. Norton.

Lawrence, B., and W. H. Bossert. 1969. "The Cranial Evidence for Hybridization in New England *Canis.*" *Museum of Comparative Zoology Harvard University Brevoria* 330:1–13.

Chapter 1

Arnold, M. L. 1997. *Natural Hybridization and Evolution.* New York: Oxford University Press.

Avise, J. C. 1994. *Molecular Markers, Natural History, and Evolution.* New York: Chapman and Hall.

———. 2000. *Phylogeography: The History and Formation of Species*. Cambridge, MA: Harvard University Press.

Barluenga, M., K. N Stölting, W. Salzburger, M. Muschick, and A. Meyer. 2006. "Sympatric Speciation in Nicaraguan Crater Lake Cichlid Fish." *Nature* 439:719–23.

Beck, A. M. 1975. "The Ecology of 'Feral' and Free Roving Dogs in Baltimore." In *The Wild Canids*, ed. M. W. Fox, 380–90. New York: Van Nostrand Reinhold Co.

Berlocher, S. H., and J. L. Feder. 2002. "Sympatric Speciation in Phytophagous Insects: Moving beyond Controversy." *Annual Review of Entomology* 47:773–815.

Berman, M., and I. Dunbar. 1983. "The Social Behaviour of Free-Ranging Suburban Dogs." *Applied Animal Ethology* 10:5–17.

Coppinger, R., and L. Coppinger. 2001. *Dogs: A New Understanding of Canine Origin, Behavior, and Evolution*. New York: Scribner.

Coppinger, R., L. Spector, and L. Miller. 2009. "What, If Anything, Is a Wolf?" In *The World of Wolves: New Perspectives on Ecology, Behaviour and Management*, ed. M. Musiani, L. Boitani, and P. C. Paquet, 41–67. Calgary: University of Calgary Press.

Drake, A. G. 2004. "Evolution and Development of the Skull Morphology of Canids: An Investigation of Morphological Integration and Heterochrony." PhD Diss., University of Massachusetts, Amherst.

Fielding, W. J., J. Mather, and M. Issacs. 2005. *Potcakes: Dog Ownership in New Providence, the Bahamas*. West Lafayette, IN: Purdue University Press.

Gould, S. J. 2002. *The Structure of Evolutionary Theory*. Cambridge, MA: Harvard University Press.

Larson, G., E. K. Karlsson, A. Perri, M. T. Webster, S. Y. W. Ho, and J. Peters, et al. 2012. "Rethinking Dog Domestication by Integrating Genetics, Archeology, and Biogeography." *Proceedings of the National Academy of Sciences of the United States of America* 109:8878–83.

Wang, X., and R. H. Tedford. 2008. *Dogs: Their Fossil Relatives and Evolutionary History*. New York: Columbia University Press.

Wayne, R. K. 1986. "Cranial Morphology of Domestic and Wild Canids: The Influence of Development on Morphological Change." *Evolution* 40:243–61.

Wood, A. E. 1957. "What, If Anything, Is a Rabbit?" *Evolution* 11:417–25.

Zimen, E. 1981. *The Wolf, a Species in Danger*. New York: Delacorte Press.

Chapter 2

Beck, A. M. 1973. *The Ecology of Stray Dogs: A Study of Free-Ranging Urban Animals*. Baltimore, MD: York Press.

Boitani, L., P. Ciucci, and A. Ortolani. 2005. "Behaviour and Social Ecology of Free-Ranging Dogs." In *The Behavioural Biology of Dogs*, ed. P. Jensen, 147–65. Oxfordshire: CABI.

Boitani, L., F. Francisci, P. Ciucci, and G. Andreoli. 1995. "Population Biology and Ecology of Feral Dogs in Italy." In *The Domestic Dog: Its Evolution, Behavior, and Interactions with People,* ed. J. A. Serpell, 218–44. Cambridge: Cambridge University Press.

Cleaveland, S., E. M. Fevre, M. Kaare, and P. G. Coleman. 2002. "Estimating Human Rabies Mortality in the United Republic of Tanzania from Dog Bite Injuries." *Bulletin of the World Health Organization* 80:304–10.

Coleman, P. G., and C. Dye. 1996. "Immunization Coverage Required to Prevent Outbreaks of Dog Rabies." *Vaccine* 14:185–86.

Daniels, T. J., and M. Bekoff. 1989. "Populations and Social Biology of Free-Ranging Dogs, *Canis familiaris.*" *Journal of Mammalogy* 70:754–62.

Fuller, T. K., L. D. Mech, and J. F. Cochrane. 2003. "Wolf Population Dynamics." In *Wolves: Behavior, Ecology, and Conservation,* ed. L. D. Mech and L. Boitani, 161–91. Chicago: University of Chicago Press.

Jackman, J., and A. Rowan. 2007. "Free-Roaming Dogs in Developing Countries: The Public Health and Animal Welfare Benefits of Capture, Neuter, and Return Programs." In *State of the Animals,* ed. Deborah Salem and Andrew Rowan, 55–78. Washington, DC: Humane Society Press.

Lord, K., M. Feinstein, B. Smith, and R. Coppinger. 2013. "Variation in Reproductive Traits of Members of the Genus *Canis* with Special Attention to the Domestic Dog (*Canis familiaris*)." *Behavioural Processes* 92:131–42.

Lorenz, J., R. Coppinger, and M. Sutherland. 1986. "Causes and Economic Effects of Mortality in Livestock Guarding Dogs." *Journal of Range Management* 39:293–95.

Marker, L., A. J. Dickman, and D. W. Macdonald. 2005. "Survivorship and Causes of Mortality for Livestock-Guarding Dogs on Namibian Rangeland." *Rangeland Ecology and Management* 58:337–43.

Oppenheimer, C., and J. R. Oppenheimer. 1975. "Certain Behavioral Features in the Pariah Dog (*Canis familiaris*) in West Bengal." *Applied Animal Ethology* 2:81–92.

Ortolani, A., and R. P. Coppinger. 2005. "African Village Dogs: Behaviour, Ecology and Human Interactions of Free-Ranging Dogs in Ethiopia." Paper presented at the International Ethology Conference, August 22–26, Budapest, Hungary.

Ortolani, A., H. Vernoij, and R. Coppinger. 2009. "Ethiopian Village Dogs: Behavioural Responses to a Stranger's Approach." *Applied Animal Behaviour Science* 119: 210–18.

Price, E. O. 1984. "Behavioral Aspects of Animal Domestication." *Quarterly Review of Biology* 59:1–32.

World Health Organization. 1988. *Report of WHO Consultation on Dog Ecology Studies Related to Rabies Control, Geneva, 22–25 February.* Geneva: World Health Organization.

———. 2000. "RabNet: Human and Animal Rabies. Country: Ethiopia." http://www.who.int/rabies/rabnet/en. (Site discontinued.)

Chapter 3

Alberch, P., S. J. Gould, G. F. Oster, and D. B.Wake. 1979. "Size and Shape in Ontogeny and Phylogeny." *Paleobiology* 5:296–31.

Coppinger, R., and L. Coppinger. 2001. *Dogs: A New Understanding of Canine Origin, Behavior, and Evolution.* New York: Scribner.

Darwin, C. 1858. "On the Tendency of Species to Form Varieties and on the Perpetuation of Varieties of Species by Natural Selection." *Journal of the Linnean Society of London Zoology* 3:45–62.

Drake, A. G., M. Coquerelle, and G. Colombeau. 2015. "3D Morphometric Analysis of Fossil Canid Skulls Contradicts the Suggested Domestication of Dogs during the Late Paleolithic." *Scientific Reports,* 5. http://www.nature.com/srep/2015/150205/srep08299/full/srep08299.html.

Fox, M. W. 1978. *The Dog: Its Domestication and Behavior.* New York: Garland STPM Press.

Geist, V. 1971. *Mountain Sheep: A Study in Behavior and Evolution.* Chicago: University of Chicago Press.

———. 1987. "On Speciation in Ice Age Mammals, with Special Reference to Cervids and Caprids." *Canadian Journal of Zoology* 65:1067–84.

———. 1998. *Deer of the World: Their Evolution, Behavior and Ecology.* Mechanicsburg, PA: Stackpole Books.

Germonpré, M., et al. 2009. "Fossil Dogs and Wolves from Paleolithic Sites in Belgium, the Ukraine and Russia: Osteometry, Ancient DNA and Stable Isotopes." *Journal of Archaeological Science* 36:473–90.

Honacki, J. H., K. E. Kinman, and J. W. Koeppl, eds. 1982. *Mammal Species of the World: A Taxonomic and Geographic Reference.* Lawrence, KS: Allen Press and Association of Systematic Collections.

Larson, G., E. K. Karlsson, A. Perri, M. T. Webster, S. Y. W. Ho, and J. Peters, et al. 2012. "Rethinking Dog Domestication by Integrating Genetics, Archeology, and Biogeography." *Proceedings of the National Academy of Sciences of the United States of America* 109:8878–83.

Lorenz. K. 1953. *Man Meets Dog.* Boston: Houghton Mifflin Company.

Morey, D. F. 1992. "Size, Shape, and Development in the Evolution of the Domestic Dog." *Journal of Archaeological Science* 19:181–204.

———. 1994. "The Early Evolution of the Domestic Dog." *American Scientist* 82:336–47.

Olsen, S. J., and J. W. Olsen, J.W. 1977. "The Chinese Wolf, Ancestor of New World Dogs." *Science* 197, no. 4303: 533–35.

Ortolani, A., H. Vernoij, and R. Coppinger. 2009. "Ethiopian Village Dogs: Behavioural Responses to a Stranger's Approach." *Applied Animal Behaviour Science* 119:210–18.

Sands, M., R. P. Coppinger, and C. J. Phillips. 1977 "A Comparison of Thermal

Sweating and Histology of Sweat Glands of Selected Canids." *Journal of Mammalogy* 58:74–78.

Vilà, C., P. Savolainen, J. E. Maldonado, I. R. Amorim, J. E. Rice, R. L. Honeycutt, K. A. Crandall, J. Lundeberg, and R. K. Wayne. 1997. "Multiple and Ancient Origins of the Domestic Dog." *Science* 276, no. 5319: 1687–89.

Chapter 4

Boitani, L., F. Francisci, P. Ciucci, and G. Andreoli. 1995. "Population Biology and Ecology of Feral Dogs in Italy." In *The Domestic Dog: Its Evolution, Behavior, and Interactions with People*, ed. J. A. Serpell, 218–44. Cambridge: Cambridge University Press.

Coppinger, R., and L. Coppinger. 2001. *Dogs: A New Understanding of Canine Origin, Behavior and Evolution*. New York: Scribner.

Darwin, C. 1858. "On the Tendency of Species to Form Varieties and on the Perpetuation of Varieties of Species by Natural Selection." *Journal of the Linnean Society of London Zoology* 3:45–62.

———. 1899. *The Variation of Animals and Plants under Domestication*. Vol. 1. New York: Appleton.

Geist, V. 1971. *Mountain Sheep: A Study in Behavior and Evolution*. Chicago: University of Chicago Press.

———. 1987. "On Speciation in Ice Age Mammals, with Special Reference to Cervids and Caprids." *Canadian Journal of Zoology* 65:1067–84.

Gottelli, D., C. Sillero-Zubiri, G. D. Applebaum, M. S. Roy, D. J. Girman, J. Garcia-Moreno, et al. 1994. "Molecular Genetics of the Most Endangered Canid: The Ethiopian Wolf, *Canis simensis*." *Molecular Ecology* 3:301–31.

Haldane, J. B. S. 1930. *The Causes of Evolution*. London: Longmans, Green.

Honacki, J. H., K. E. Kinman, and J. W. Koeppl, eds. 1982. *Mammal Species of the World: A Taxonomic and Geographic Reference*. Lawrence, KS: Allen Press and Association of Systematic Collections.

Lewontin, R. C. 1978. "Adaptation." *Scientific American* 239, no. 9:156–59.

Macdonald, D., and G. Carr. 1995. "Variation in Dog Society: Between Resource Dispersion and Social Flux." In *The Domestic Dog: Its Evolution, Behaviour and Interactions with People*, ed. J. Serpell, 199–216. Cambridge: Cambridge University Press.

Macdonald, D. W., S. Creel, and M. G. L. Millis. 2004. "Canid Society." In *Biology and Conservation of Wild Animals*, ed. D. W. Macdonald, and C. Sillero-Zuberi, 85–106. Oxford: Oxford University Press.

Ortolani, A., and R. P. Coppinger. 2005. "African Village Dogs: Behaviour, Ecology and Human Interactions of Free-Ranging Dogs in Ethiopia." Paper presented at the International Ethology Conference, August 22–26, Budapest, Hungary.

Smith, D. W., D. R. Stahler, E. Stahler, M. Metz, K. Quimby, R. McIntyre, C. Ruhl, H. Martin, R. Kindermann, N. Bowersock, and M. McDevitt. 2013. *Yellowstone*

Wolf Project: Annual Report, 2012. YCR-2013–02. Yellowstone National Park, WY: National Park Service, Yellowstone Center for Resources.

Vucetich, J. A., R. O. Peterson, and M. P. Nelson. 2010. "Will the Future of Isle Royale Wolves and Moose Always Differ from Our Sense of Their Past? In *The World of Wolves: New Perspectives on Ecology, Behaviour and Policy*, ed. M. Musiani, L. Boitani, and P. Paquet, 123–154. Calgary: University of Calgary Press.

Chapter 5

Andelt, W. F. 1985. "Behavioral Ecology of Coyotes in South Texas." *Wildlife Monographs* 94:3–45.

Boitani, L., F. Francisci, P. Ciucci, and G. Andreoli. 1995. "Population Biology and Ecology of Feral Dogs in Italy." In *The Domestic Dog: Its Evolution, Behavior, and Interactions with People*, ed. J. A. Serpell, 218–44. Cambridge: Cambridge University Press.

Macdonald, D. W., S. Creel, and M. G. L. Millis. 2004. "Canid Society." In *Biology and Conservation of Wild Animals*, ed. D. W. Macdonald, and C. Sillero-Zuberi, 85–106. Oxford: Oxford University Press.

Macdonald, D. W., and P. D. Moehlman. 1982. "Cooperation, Altruism, and Restraint in the Reproduction of Carnivores." *Perspectives in Ethology* 5:433–67.

Mech, L. D. 1995. "Summer Movements and Behavior of an Arctic Wolf *Canis lupus*, Pack without Pups." *Canadian Field-Naturalist* 110:473–75.

Mech, L. D., and L. Boitani. 2003. "Wolf Social Ecology." In *Wolves: Behavior, Ecology, and Conservation*, ed. L. D. Mech, and L. Boitani, 1–34. Chicago: University of Chicago Press.

Moehlman, P. D. 1983. "Socioecology of Silverbacked and Golden Jackals (*Canis mesomelas* and *Canis aureus*)." In *Recent Advances in the Study of Mammalian Behavior*, ed. J. F. Eisenberg and D. G. Kleiman, 423–53. Lawrence, KS: American Society of Mammalogists.

Smith, D. W., D. R. Stahler, E. Stahler, M. Metz, K. Quimby, R. McIntyre, C. Ruhl, H. Martin, R. Kindermann, N. Bowersock, and M. McDevitt. 2013. *Yellowstone Wolf Project: Annual Report, 2012*. YCR-2013–02. Yellowstone National Park, WY: National Park Service, Yellowstone Center for Resources.

Vucetich, J. A., and R. Peterson. 2014. *Ecological Studies of Wolves on Isle Royale: Annual Report, 2013–14*. Houghton, MI: School of Forest Resources and Environmental Science, Michigan Technological University.

Vucetich, J. A., R. O. Peterson, and M. P. Nelson. 2010. "Will the Future of Isle Royale Wolves and Moose Always Differ from Our Sense of Their Past? In *The World of Wolves: New Perspectives on Ecology, Behaviour and Policy*, ed. M. Musiani, L. Boitani, and P. Paquet, 123–154. Calgary: University of Calgary Press.

Chapter 6

Coppinger, R., and L. Coppinger. 2001. *Dogs: A New Understanding of Canine Origin, Behavior and Evolution.* New York: Scribner.

European Pet Food Industry Federation. 2010. *Facts and Figures.* Brussels, Belgium. http://www.fediaf.org/fileadmin/user_upload/facts_and_figures_2010.pdf.

Medina, M. 2007. *The World's Scavengers: Salvaging for Sustainable Consumption and Production.* New York: Altamira Press.

Chapter 7

Boitani, L., F. Francisci, P. Ciucci, and G. Andreoli. 1995. "Population Biology and Ecology of Feral Dogs in Italy." In *The Domestic Dog: Its Evolution, Behavior, and Interactions with People,* ed. J. A. Serpell, 218–44. Cambridge: Cambridge University Press.

Cafazzo, S., P. Valsecchi, R. Bonanni, and F. Natoli. 2010. "Dominance in Relation to Age, Sex, and Competitive Contexts in a Group of Free-Ranging Domestic Dogs." *Behavioral Ecology* 21, no. 3: 443–55.

Creel, S., J. A. Winnie, D. Christianson, and S. Lily. 2008. "Time and Space in General Models of Antipredator Response: Tests with Wolves and Elk." *Animal Behavior* 76:1139–46.

Darwin, C. 1858. "On the Tendency of Species to Form Varieties and on the Perpetuation of Varieties of Species by Natural Selection." *Journal of the Linnean Society of London Zoology* 3:45–62.

European Pet Food Industry Federation. 2010. *Facts and Figures.* Brussels, Belgium. http://www.fediaf.org/fileadmin/user_upload/facts_and_figures_2010.pdf.

Lorenz, J. R. 1976. "Characteristics of Massachusetts and Vermont Coyotes." *Northern Raven: The Quarterly Newsletter of the Center for Northern Studies* 4, nos. 3–4: 11–13.

MacNulty, D. R., D. W. Smith, L. D. Mech, and L. E. Eberly. 2009. "Body Size and Predatory Performance in Wolves: Is Bigger Better?" *Journal of Animal Ecology* 78:532–39.

Muro, C., R. Escobedo, L. Spector, and R. P. Coppinger. 2011. "Wolf-Pack (*Canis lupus*) Hunting Strategies Emerge from Simple Rules in Computational Simulations." *Behavioural Processes* 88:192–97.

Oaks, J. L., M. Gilbert, M. Z. Virani, R. T. Watson, C. U. Meteyer, B. A. Rideout, H. L. Shivaprasad, A. Shakeel, J. I. C. Muhammad, A. Muhammad, M. Shahid, A. Ahmad, and A. K. Aleem. 2004. "Diclofenac Residues as the Cause of Vulture Population Decline in Pakistan." *Nature* 427:630.

Peterson, R. O., N. J. Thomas, J. M. Thurber, J. A. Vucetich, and T. A. Waite. 1998. "Population Limitation and the Wolves of Isle Royale." *Journal of Mammalogy* 79:828–41.

Peterson, R. O., J. D. Woolington, and T. N. Bailey. 1984. "Wolves of the Kenai Peninsula, Alaska." *Wildlife Monographs* 88:1–52.

Schmidt, P. A., and L. D. Mech. 1997. "Wolf Pack Size and Food Acquisition." *American Naturalist* 150:513–17.

Smith, D. W., D. R. Stahler, E. E. Albers, M. Metz, K. Cassidy, J. Irving, R. Raymond, C. Anton, N. Bowersock, H. Zaranek, and R. McIntyre. 2010. *Yellowstone Wolf Project: Annual Report, 2009.* YCR-2009–01. Yellowstone National Park, WY: National Park Service, Yellowstone Center for Resources.

Vucetich, J. A., R. O. Peterson, and T. A. Waite. 2004. "Raven Scavenging Favors Group Foraging in Wolves." *Animal Behaviour* 67:1117–26.

Vucetich, J. A., D. W. Smith, and D. R. Stahler. 2005. "Influence of Harvest, Climate and Wolf Predation on Yellowstone Elk, 1961–2004." *Oikos* 111:259–70.

Will, G. B., and J. R. Lorenz. 1980. "Movement of an Eastern Coyote from Vermont to New York." *New York Fish and Game Journal* 27, no. 1: 94–95.

Chapter 8

Andelt, W. F., D. P. Althoff, and P. S. Gipson. 1979. "Movements of Breeding Coyotes with Emphasis on Den Site Relationships." *Journal of Mammalogy* 60:568–75.

Atkinson, R. P., and A. J. Loveridge. 2004. "Side-Striped Jackal *Canis adustus* Sundevall, 1847." In *Canids: Foxes, Wolves, Jackals and Dogs,* ed. C. Sillero-Zubiri, M. Hoffmann, and D. W. Macdonald, 152–56. Gland, Switzerland: International Union for Conservation of Nature and Natural Resources.

Clutton-Brock, T. H. 2002. "Behavioral Ecology—Breeding Together: Kin Selection and Mutualism in Cooperative Vertebrates." *Science* 296:69–72.

Clutton-Brock, T. H., A. F. Russell, L. L. Sharpe, P. N. M. Brotherton, G. M. McIlrath, S. White, and E. Z. Cameron,. 2001. "Effects of Helpers on Juvenile Development and Survival in Meerkats." *Science* 293:2446–49.

Creel, N. 2001. "Interspecific Competition and the Population Biology of Extinction-Prone Carnivores. In *Carnivore Conservation,* ed. J. L. Gittleman, S. M. Funk, D. D. Macdonald, and R. K. Wayne, 35–60. Cambridge: Cambridge University Press.

Daniels, T. J. 1983. "The Social Organization of Free-Ranging Urban Dogs. Pt. 1, Non-estrous Social Behaviour." *Applied Animal Ethology* 10:341–63.

Geist, V. 1998. *Deer of the World: Their Evolution, Behavior and Ecology.* Mechanicsburg, PA: Stackpole Books.

Ghosh, B., D. K. Choudhuri, and B. Pal. 1984. "Some Aspects of the Sexual Behavior of Stray Dogs, *Canis familiaris.*" *Applied Animal Behaviour Science* 13:113–27.

Lord, K., M. Feinstein, B. Smith, and R. Coppinger. 2013. "Variation in Reproductive Traits of Members of the Genus *Canis* with Special Attention to the Domestic Dog (*Canis familiaris*)." *Behavioural Processes* 92:131–42.

MacNulty, D. R., D. W. Smith, J. A. Vucetich, L. D. Mech, D. R. Stahler, and C. Packer. 2009. "Predatory Senescence in Ageing Wolves." *Ecology Letters* 12:1–10.

Mech, L. 2002. "Breeding Season of Wolves, *Canis lupus*, in Relation to Latitude." *Canadian Field-Naturalist* 116:139–40.

Mech, L. D., and M. E. Nelson. 1989. "Polygyny in a Wild Wolf Pack." *Journal of Mammalogy* 70:675–76.

Mech, L. D., P. C. Wolf, and J. M. Packard. 1999. "Regurgitative Food Transfer among Wild Wolves." *Canadian Journal of Zoology* 77:1192–95.

Medjo, D. C., and L. D. Mech. 1976. "Reproductive Activity in Nine- and Ten-Month-Old Wolves." *Journal of Mammalogy* 57:406–8.

Moehlman, P. D. 1979. "Jackal Helpers and Pup Survival." *Nature* 277:382–83.

Pal, S. K., B. Ghosh, and S. Roy. 1998. "Agonistic Behaviour of Free-Ranging Dogs (*Canis familiaris*) in Relation to Season, Sex and Age." *Applied Animal Behaviour Science* 59:331–48.

———. 1998. "Dispersal Behaviour of Free-Ranging Dogs (*Canis familiaris*) in Relation to Age, Sex, Season and Dispersal Distance." *Applied Animal Behaviour Science* 61:123–32.

———. 1999. "Inter- and Intra-Sexual Behaviour of Free-Ranging Dogs (*Canis familiaris*)." *Applied Animal Behaviour Science* 62:267–78.

Chapter 9

Mosser, A., and C. Packer. 2009. "Group Territoriality and the Benefits of Sociality in the African Lion, *Panthera leo*." *Animal Behaviour* 78:359–70.

Ortolani, A., H. Vernoij, and R. Coppinger. 2009. "Ethiopian Village Dogs: Behavioural Responses to a Stranger's Approach." *Applied Animal Behaviour Science* 119:210–18.

Chapter 10

Boitani, L., P. Ciucci, and A. Ortolani. 2005. "Behaviour and Social Ecology of Free-Ranging Dogs." In *The Behavioural Biology of Dogs*, ed. P. Jensen, 147–65. Oxfordshire: CABI.

Boitani, L., F. Francisci, P. Ciucci, and G. Andreoli. 1995. "Population Biology and Ecology of Feral Dogs in Italy." In *The Domestic Dog: Its Evolution, Behavior, and Interactions with People*, ed. J. A. Serpell, 218–44. Cambridge: Cambridge University Press.

Butler, J. R. A., J. T. duToit, and J. Bingham. 2004. "Free-Ranging Domestic Dogs (*Canis familiaris*) as Predators and Prey in Rural Zimbabwe: Threats of Competition and Disease to Large Wild Carnivores. *Biological Conservation* 115:369–78.

Coppinger, R. P., and C. K. Smith. 1983. "The Domestication of Evolution." *Environmental Conservation* 10:283–92.

———. 1990. "A Model for Understanding the Evolution of Mammalian Behavior." In *Current Mammalogy*, ed. Hugh Genoways, 2:335–74. New York: Plenum Press.

Frank, H., and M. G. Frank. 1982. "On the Effects of Domestication on Canine Social Development and Behavior." *Applied Animal Ethology* 8:507–25.

Gallant, J. 2002. *The Story of the African Dog*. Scottsville, South Africa: University of KwaZulu-Natal Press.

Lankester, F., K. Hampson, T. Lembo, G. Palmer, L. Taylor, and S. Cleaveland. 2014. "Implementing Pasteur's Vision for Rabies Elimination." *Science* 345, no. 6204: 1562–64.

Lantis, M. 1980. "Changes in the Alaskan Eskimo Relation of Man to Dog and Their Effect on Two Human Diseases." *Arctic Anthropology* 17:2–24.

Lord, K., M. Feinstein, and R. Coppinger. 2009. "Barking and Mobbing." *Behavioural Processes* 81:358–68.

Medina, M. 2007. *The World's Scavengers: Salvaging for Sustainable Consumption and Production*. New York: Altamira Press.

Ortolani, A., H. Vernoij, and R. Coppinger. 2009. "Ethiopian Village Dogs: Behavioural Responses to a Stranger's Approach." *Applied Animal Behaviour Science* 119:210–18.

Pellis, S. M., and V. C. Pellis. 1996. "On Knowing It's Only Play: The Role of Play Signals in Play Fighting." *Aggression and Violent Behavior* 1:249–68.

Rindos, D. 1980. "Symbiosis, Instability and the Origins and Spread of Agriculture: A New Model." *Current Anthropology* 21:751–72.

Ruiz-Izaguirre, E., and C. H. A. M. Eilers. 2012. "Perceptions of Village Dogs by Villagers and Tourists in the Coastal Region of Rural Oaxaca, Mexico." *Anthrozoös* 25:75–91.

Thomson, K. S. 1996. "The Fall and Rise of the English Bulldog." *American Scientist* 84:220–23.

Varner, J. G., and J. J. Varner. 1983. *Dogs of the Conquest*. Norman: University of Oklahoma Press.

Chapter 11

Ruiz-Izaguirre, E., and C. H. A. M. Eilers. 2012. "Perceptions of Village Dogs by Villagers and Tourists in the Coastal Region of Rural Oaxaca, Mexico." *Anthrozoös* 25:75–91.

Lord, K., 2013. "A Comparison of the Sensory Development of Wolves (*Canis lupus lupus*) and Dogs (*Canis lupus familiaris*)." *Ethology* 119:110–20.

Serpell, J. 1986. *In the Company of Animals*. Oxford: Basil Blackwell.

Chapter 12

Coppinger, R., and L. Coppinger. 2001. *Dogs: A New Understanding of Canine Origin, Behavior and Evolution*. New York: Scribner.

Drake, A. G., and C. P. Klingenberg. 2008. "The Pace of Morphological Change: Historical Transformation of Skull Shape in St Bernard Dogs." *Proceedings of the Royal Society B: Biological Sciences* 275:71–76.

Fentress, J. C. 1967. "Observations on the Behavioral Development of a Hand-Reared Male Timber Wolf." *American Zoologist* 7:339–51.

Price, E. O. 1998. "Behavioral Genetics and the Process of Animal Domestication."

In *Genetics and the Behavior of Domestic Animals*, ed. T. Grandin, 31–65. San Diego, CA: Academic Press.

Serpell, J. 1986. *In the Company of Animals*. Oxford: Basil Blackwell.

Wilson, E. O. 1984. *Biophilia*. Cambridge, MA: Harvard University Press.

Chapter 13

Beck, A. M. 1973. *The Ecology of Stray Dogs: A Study of Free-Ranging Urban Animals*. Baltimore, MD: York Press.

Coppinger, R., J. Lorenz, J. Glendinning, and P. Pinardi. 1983. "Attentiveness of Guarding Dogs for Reducing Predation on Domestic Sheep." *Journal of Range Management* 36, no. 3: 275–79.

Drake, A. G., and C. P. Klingenberg. 2008. "The Pace of Morphological Change: Historical Transformation of Skull Shape in St Bernard Dogs." *Proceedings of the Royal Society B: Biological Sciences* 275:71–76.

Gallant, J., and E. Gallant. 2008. *SOS Dog: The Purebred Dog Hobby Re-examined*. Crawford CO: Alpine Publications.

Price, E. O. 1998. "Behavioral Genetics and the Process of Animal Domestication." In *Genetics and the Behavior of Domestic Animals*, ed. T. Grandin, 31–65. San Diego, CA: Academic Press.

Thomson, K. S. 1996. "The Fall and Rise of the English Bulldog." *American Scientist* 84:220–23.

Varner, J. G., and J. J. Varner. 1983. *Dogs of the Conquest*. Norman: University of Oklahoma Press.

Chapter 14

Beck, A. M. 1973. *The Ecology of Stray Dogs: A Study of Free-Ranging Urban Animals*. Baltimore, MD: York Press.

Boitani, L. 1983. "Wolf and Dog Competition in Italy." *Acta Zoologica Fennica* 174:259–64.

Boitani, L., F. Francisci, P. Ciucci, and G. Andreoli. 1995. "Population Biology and Ecology of Feral Dogs in Italy." In *The Domestic Dog: Its Evolution, Behavior, and Interactions with People*, ed. J. A. Serpell, 218–44. Cambridge: Cambridge University Press.

Brisbin, I. L., R. P. Coppinger, M. H. Feinstein, S. N. Austad, and J. J. Mayer, J. J. 1994. "The New Guinea Singing Dog: Taxonomy, Captive Studies and Conservation Priorities." *Science in New Guinea* 20:27–38.

Coppinger, R. P., C. K. Smith, and L. Miller. 1985. "Observations on Why Mongrels May Make Effective Livestock Protecting Dogs." *Journal of Range Management* 38, no. 6: 560–61.

Fox, M. W. 1978. *The Dog: Its Domestication and Behavior*. New York: Garland STPM Press.

Jackman, J., and A. Rowan. 2007. "Free-Roaming Dogs in Developing Countries: The Public Health and Animal Welfare Benefits of Capture, Neuter, and Return

Programs." In *State of the Animals*, ed. Deborah Salem and Andrew Rowan, 55–78. Washington, DC: Humane Society Press.

Macdonald, D., and G. Carr. 1995. "Variation in Dog Society: Between Resource Dispersion and Social Flux." In *The Domestic Dog: Its Evolution, Behaviour and Interactions with People*, ed. J. Serpell, 199–216. Cambridge: Cambridge University Press.

Wayne, R. K., N. Lehman, and T. K. Fuller. 1995. "Conservation Genetics of the Gray Wolf." In *Ecology and Conservation of Wolves in a Changing World*, ed. L. N. Carbyn, S. H. Fritts, and D. R. Siep, 399–407. Occasional Publication no. 35. Edmonton: Canadian Circumpolar Institute.

Young, S. P., and Goldman, E. A. 1944. *The Wolves of North America*. Washington, DC: America Wildlife Institute.

Chapter 15

Coppinger, R., L. Spector, and L. Miller. 2009. "What, If Anything, Is a Wolf?" In *The World of Wolves: New Perspectives on Ecology, Behaviour and Management*, ed. M. Musiani, L. Boitani, and P. C. Paquet, 41–67. Calgary: University of Calgary Press.

Godinho, R., L. Llaneza, J. C. Blanco, S. Lopes, F. Alvares, E. J. García, V. Palacios, Y. Cortés, J. Talegón, and N. Ferrand. 2011. "Genetic Evidence for Multiple Events of Hybridization between Wolves and Domestic Dogs in the Iberian Peninsula." *Molecular Ecology* 20:5154–66.

Gottelli, D., C. Sillero-Zubiri, G. D. Applebaum, M. S. Roy, D. J. Girman, J. Garcia-Moreno, et al. 1994. "Molecular Genetics of the Most Endangered Canid: The Ethiopian Wolf, *Canis simensis*." *Molecular Ecology* 3:301–31.

Gould, S. J. 2002. *The Structure of Evolutionary Theory*. Cambridge, MA: Harvard University Press.

Hindrikson, M., P. Männil, J. Ozolins, A. Krzywinski, and U. Saarma. 2012. "Bucking the Trend in Wolf-Dog Hybridization: First Evidence from Europe of Hybridization between Female Dogs and Male Wolves." *PLoS One* 7, no. 10: e46465.

Howard, W. E. 1949. "A Means to Distinguish Skull of Coyotes and Domestic Dogs." *Journal of Mammalogy* 30:169–71.

Larson, G., E. K. Karlsson, A. Perri, M. T. Webster, S. Y. W. Ho, and J. Peters, et al. 2012. "Rethinking Dog Domestication by Integrating Genetics, Archeology, and Biogeography." *Proceedings of the National Academy of Sciences of the United States of America* 109:8878–83.

Stebbins, G. L. 1959. "The Role of Hybridization in Evolution." *Proceedings of the American Philosophical Society* 103:231–51.

Vilà, C., C. Walker, A.-K. Sundqvist, Ø. Flagstad, Z. Andersone, A. Casulli, I. Kojola, H. Valdmann, J. Halverson, and H. Ellegren. 2003. "Combined Use of Maternal, Paternal and Bi-Parental Genetic Markers for the Identification of Wolf-Dog Hybrids." *Heredity* 90:17–24.

Wilson, P. J., L. Y. Rutledge, T. J. Wheeldon, B. R. Patterson, and B. N. White. 2012.

"Y-Chromosome Evidence Supports Widespread Signatures of Three-Species *Canis* Hybridization in Eastern North America." *Ecology and Evolution* 2, no. 9: 2325–32.

Chapter 16

Boitani, L., F. Francisci, P. Ciucci, and G. Andreoli. 1995. "Population Biology and Ecology of Feral Dogs in Italy." In *The Domestic Dog: Its Evolution, Behavior, and Interactions with People*, ed. J. A. Serpell, 218–44. Cambridge: Cambridge University Press.

Clutton-Brock, J., ed. 1989. *The Walking Larder: Patterns of Domestication, Pastoralism, and Predation*. Boston: Unwin Hyman.

Clutton-Brock, J., G. B. Corbet, and M. Hills. 1976. "A Review of the Family Canidae, with a Classification by Numerical Methods." *Bulletin of the British Museum (Natural History) Zoology* 29:117–99.

Drake, A. G. 2011. "Dispelling Dog Dogma: An Investigation of Heterochrony in Dogs Using 3D Geometric Morphometric Analysis of Skull Shape." *Evolution and Development* 13, no. 2:204–13.

Hole, F., and C. Wyllie. 2007. "The Oldest Depictions of Canines and a Possible Early Breed of Dog." *Paléorient* 33, no. 1: 175–85.

Swart, S. 2003. "Dogs and Dogma: A Discussion of the Socio-Political Construction of Southern African Dog 'Breeds' as a Window into Social History." *South African Historical Journal* 48:4–31.

Zeuner, F. E. 1963. *A History of Domesticated Animals*. New York: Harper and Row.

Chapter 17

Drake, A. G., M. Coquerelle, and G. Colombeau. 2015. "3D Morphometric Analysis of Fossil Canid Skulls Contradicts the Suggested Domestication of Dogs during the Late Paleolithic." *Scientific Reports*, 5. http://www.nature.com/srep/2015/150205/srep08299/full/srep08299.html.

Drake, A. G., and C. P. Klingenberg. 2010. "Large-Scale Diversification of Skull Shape in Domestic Dogs: Disparity and Modularity." *American Naturalist* 175:289–301.

Jackman, J., and A. Rowan. 2007. "Free-Roaming Dogs in Developing Countries: The Public Health and Animal Welfare Benefits of Capture, Neuter, and Return Programs." In *State of the Animals*, ed. Deborah Salem and Andrew Rowan, 55–78. Washington, DC: Humane Society Press.

Price, E. O. 1998. "Behavioral Genetics and the Process of Animal Domestication." In *Genetics and the Behavior of Domestic Animals*, ed. T. Grandin, 31–65. San Diego, CA: Academic Press.

INDEX

abattoirs, 93–94, 96, 154
acquisition cost, 80, 81; of dog vs. wolf, 98; of scavengers on people, 90, 92, 95
adaptation, and behavioral ecology, 66
adaptation to a niche: benefit/cost ratio and, 68; concept of, 44, 47, 68; dog produced by, 42, 68; specific dog environments and, 172. *See also* natural selection; niche
adoption of animals by dogs, 159–64
adoption of dogs by people: by children, 159, 161, 162, 167–69; by programs capturing village dogs, 227–28
adoption of people by dogs, 119, 155–65
African dog, xvi, 158
age structure of populations, 27–30; Boitani's findings on, 30, 144; of Mexico City dump dogs, 29–30, 119, 156; over-reproduction and, 105–6; sterilization plans and, 226;

survival of juvenile dogs and, 119; vaccination plans and, 29, 30, 143–44; of wolves, 28, 67
agriculture: creating new niches for many species, 45, 220–21; creating niche for village dogs, 43, 219
akbash, 52–53, 182
Alexander the Great, 181–82
Algonquin wolf, xiv, 4
altitude and latitude, adaptation to, 36–37, 212
amensalism, 131
Anatolian shepherds, 12, 52–53, 146, 179, 180, 182
ancestry of the dog. *See* evolution of the dog
ancient breeds, mythological, 183
Aquinas, Thomas, 15
Arabian wolf, 207
Arctic wolf (*Canis lupus arctos*), 36
Aristotle, on species as essence, 14–15, 40